Proximity and Preference

Proximity and Preference

Problems in the Multidimensional Analysis of Large Data Sets

Reginald G. Golledge
University of California, Santa Barbara

John N. Rayner
Ohio State University

editors

UNIVERSITY OF MINNESOTA PRESS ☐ MINNEAPOLIS

Publication of this work has been
made possible in part by a grant from
the Andrew W. Mellon Foundation.

Published by the University of Minnesota Press,
2037 University Avenue Southeast, Minneapolis, MN 55414
Printed in the United States of America.

Library of Congress Cataloging in Publication Data
Main entry under title:

Proximity and preference.

Includes bibliographical references and index.
1. Geographical perception. 2. Space perception.
3. Multidimensional scaling. I. Golledge, Reginald G., 1937-
II. Rayner, John N.
G71.5.P76   910   81-14634
ISBN 0-8166-1042-8   AACR2

Contents

Part 3. Special Problems

Preface

Problems of dealing with large data sets have been an integral part of the life of social and behavioral scientists. It seems, in fact, that we are forever faced with one of two problems: either determining whether a given sample of data is *large enough* to be statistically significant or *reducing* a massive volume of data so as to make it at least partly comprehensible. This book is directed more to those faced with the problems of collecting and handling large data sets than to those who deal primarily with small data sets, but it does examine a problem fundamental to both groups, that of deciding just *what* data are to be collected and *how* they should be collected.

The major sources of large data sets for most social scientists are those surveys carried out by governmental authorities. These include census reports, tax statistics, land values, traffic flow, migration statistics, and so on. In many cases, the researchers are not faced with the problem of *data collection*, for some data-collecting agency has already decided on the range of variables that are to be collected, the total number of people that are to be sampled, and the way in which the data are to be presented for public consumption. More often that not, therefore, the individuals who concern themselves with already collected large data sets are faced with problems of determining a suitable sample size and checking for temporal and spatial reliability rather than solving problems of handling the volume of data collected. In recent years, however, many social and behavioral scientists have been generating their own data sets. These data sets are compiled by survey research methods and include things such as attitudinal and preference questions or other subjective evaluations that are rarely (if ever) collected by organized governmental authorities. The major question that faces such researchers is how to get the maximum amount of subjective data from respondents in such a way that its usefulness is not impaired. This question becomes even more important when various types of data other than objective counts of phenomena are desired. For example, if one wished to establish the preferences of a sample of individuals for a range of commodities, one might adopt a paired comparison data presentation format. Since the number of pairs about which judgments are required is $n(n - 1)/2$ (which increases the total number of responses very rapidly as the number of stimuli increases), the difficulty involved in obtaining a reasonable data set becomes obvious [2, 3].

In this book we present a range of papers designed to assist researchers in solving problems of designing experiments for collecting large volumes of subjective data. The data discussed in various papers in the book relate primarily to problems of *proximity* and *preference* and include stimulus pair presentation data (commonly used in psychology), preference ranking and attitudinal data for use in marketing, urban travel and migration choices, and subjective data about places for compilations of mental maps or space-preference functions.

Because the book has an interdisciplinary cast of authors, the Introduction examines various substantive and methodological aspects of the interface between geography and psychology and presents an overview of interaction between the two disciplines.

The next set of papers examines a variety of problems related to the use of large data sets with specific methodologies, ranging from multidimensional scaling through interactive computer programming to regression analysis. The papers written by psychologists tend to focus more on theoretical problems and analytical methods associated with the use of large data sets. Papers in other sections of the book concentrate on some empirical or practical problems involved in the use of various experimental designs that generate large data sets.

An exception to this general statement that the psychologists contributed the methodological papers and the others contributed the case studies appears in the paper by Spector and Rivizzigno.

These two geographers were concerned with the design of an experiment for the recovery of cognitive information about a city. Their paper describes the efficiency of cyclic and random designs as compared to an overlapping incomplete design generated arbitrarily for their specific project. Like the authors of the papers that precede theirs, however, the authors were concerned with the reliability of their data and with determining the efficiency of the experimental design that was adopted in their studies.

Much of the data that is referred to in this volume has both a spatial and a psychospatial existence. For example, Spence; Isaac; Arabie and Soli; Green; Spector and Rivizzigno; Burnett; Golledge, Rayner, and Rivizzigno; and Deutscher all discuss proximity or similarity data. Deutscher also references dominance and preference data, as do Kohler and Rushton, and Clark. Data types include paired comparisons of the proximity of places in a city, preferences for marketable stimulus objects, and preference for place and the selection of migrant destinations. Each of these problems is receiving renewed attention in disciplines such as geography, marketing, demography, economics, and sociology.

Dealing with subjective data sets of varying magnitude has been a common problem throughout the history of psychology, and this concern has spawned a particularly large volume of work on experimental design, reliability, and validity. The use of large sets of subjective data is, however, comparatively new in disciplines such as geography. Consequently, the literature on experimental design for collecting such data sets has only recently been examined in any great detail. The importance of the initial group of papers to geographers is, therefore, considerable. Such papers reference past works and provide up-to-date discussions of critical problems involved in collecting subjective data sets. Both the historical and analytical discussions contained in these papers should prove to be of considerable interest and concern to those who have to compile their own data banks of subjective evaluations of phenomena or places.

An increasing concern with subjective data relating to proximity and preference is evidenced in the range of empirical papers presented in the second and third sections of this volume, where a variety of problems relating to choice processes, estimation of preference functions, and conjoint and functional measurements are examined. Also presented are a variety of papers that illustrate various applications of multidimensional analytical methods and that discuss concepts such as comparison

of objective and subjective representations of data and data reliability problems.

As Shepard [7] points out, most of the kinds of data that are
collected in the social and behavioral sciences are not in a form
suitable for analysis by conventional metric analytical methods.
Certainly, much of the existing large data banks (such as census
materials) are exceptions to this rule, but a great variety of survey
research data are not. Many data sets are too fragmentary or too
unstable to support the determination of a well-defined latent
structure and to allow representation of results in a spatial manner. Some data sets are remarkably complete and reliable when
considered by themselves but may turn out to be incomplete or
inappropriate when considered as input for, say, a nonmetric
analytical technique such as multidimensional scaling. These
points have been emphasized by writers such as Coombs [1], Wish
[11], Green, Kehoe, and Reynolds [4], and Spence and Domoney
[10]. Shepard has also pointed out that even when data are of a
form sufficiently appropriate for use with many of the existing
nonmetric analytical methods, there is no guarantee that they
should be so used. Some data may be interpreted as being higher—
or lower—order data than they really are, and this by itself may
tend to make the data unsuitable for any rigid statistical analysis.

Considerable attention has to be paid to the nature of data
themselves, just as much attention has to be paid to the problem
of deciding *how much data* is to be collected. In an introductory
paper, Tobler outlines a range of data and measurement problems
that have plagued researchers for hundreds of years. He takes this
approach to illustrate that problems of sampling design and measurement error were as much a concern for practitioners decades
ago as they are today. He places their ideas in context by examining the problems of estimating proximity and preference that are
analyzed in the five succeeding papers.

The initial paper by Young, Sarle, and Hoffman discusses an
interactive similarity ordering system that allows a researcher to
converge toward an optimal sampling design during the course of
an experiment. The procedure can be used with a variety of data
formats (square, symmetric, asymmetric, and rectangular), and the
example used to illustrate the method uses cognitive distances
among U.S. cities. The paper by Spence investigates some of the
possible kinds of designs that can be used when a researcher collects only a fraction of the possible pairwise dissimilarity data
that could be collected. Spence's paper includes a brief review of

previous work on incomplete experimental designs and reexamines the suitability of cyclic designs for multidimensional analysis. Within the framework of his paper, he shows the results from a Monte Carlo experiment on the efficiency of cyclic designs and ties his experiments specifically into multidimensional scaling analysis. His conclusion is that highly efficient cyclic designs yield superior multidimensional scaling solutions. He also takes great pains to furnish a set of recommendations to experimenters concerning the types of designs that they should use and the most appropriate form of data that would fit both the experimental designs and their analytical methods.

Another paper on the selection of stimulus pairs for multidimensional scaling analysis is the one by Paul D. Isaac. Isaac notes that the major problem in collecting data for use with multidimensional scaling (MDS) algorithms is the rapid increase in the number of judgments required as the number of stimuli is increased. For example, designs that involve direct similarity judgments require first a solution to the problem of selecting the number of stimuli. One way of reducing the number of pairs actually judged is indicated in the paper by Spence. Isaac discusses a number of alternative strategies that may prove to have some advantages over cyclic designs in data presentation schedules. He argues that two factors should be considered in deciding how to reduce the number of pairwise judgments for a specific experiment. The first factor is the *reliability* of the particular judgments, and the second is the extent to which the particular distributions being judged are *crucial to the shape* of the configuration. Basically, therefore, the first consideration deals with the issue of the *stability* of the configuration across repeated judgments, and the second constraint deals with the *distances* most necessary to the reconstruction of the configuration. Isaac then goes ahead to comment on the degrees-of-freedom ratios that were articulated in the Spence-Domoney article [10] and discusses such ratios and the selection of designs from the point of view of error in similarity judgments.

In his paper Spence suggests that cyclic designs are the most efficient for handling large data sets; Spector and Rivizzigno test this hypothesis by considering the efficiency of cyclic and random designs against a design developed for a study in the Columbus, Ohio, area. Spector and Rivizzigno clearly show differences in the effectiveness of each experimental design with their incomplete data set and illustrate their points by commenting on the metric determinacy of configurations derived from multidimensional scaling analyses of each of the data designs. The configurations obtained are interpreted as mental maps of places in the city, and the degree of correlation between derived configurations and a set of actual locations is examined for each of the designs.

Arabie and Soli, somewhat in contrast to the preceding three papers, examine the interface between regression and methods of collecting proximity data. They suggest that, to date, a normal procedure is to tailor scaling algorithms to specific data types. They suggest that nonmetric multidimensional scaling analyses have consistently revealed types of data that yield predictable regression functions relating scaled proximities to recovered distances and argue that these observed regularities should be exploited by appropriately transforming input data so that linear regression could be used in the scaling analysis. They draw extensively on the classical Miller and Nicely data and use both two- and three-way analyses of the data. Once again, the final step is examining the interpretability of the resulting solutions and comparing such solutions with other analyses of the data that use markedly different approaches.

Hubert opens Part 2 of this volume with a discussion of the nature of preference data and highlights critical problem areas such as missing and tied data. After setting the scene for the papers that follow his, he illustrates a method of analysis somewhat different from those used by other contributors and reveals, using data taken from Clark's paper, how much results depend *both* on the quality of data and on the techniques used.

The other four papers presented in Part 2 are written by geographers and are problem oriented. The paper by Kohler and Rushton examines data designs for computing space-preference functions. The authors argue that it has never been shown that the values of space-preference functions can be accurately recovered from simulated spatial choice data. Their paper demonstrates that this is possible by applying a nonmetric multidimensional scaling method to preferences for locations. The locational stimuli are derived from simulated spatial choice data. Concentrating on preference analysis, W. A. V. Clark uses multidimensional scaling techniques on a data set of individual moves within the Milwaukee metropolitan region. The detailed nature and size of his data bank allows him to focus on movements between very small areas of the city. These small areas are considered to have characteristics (such as *stimulus content, distance* from the central business district

[CBD] , and *density*) to which the movers respond. A combination of these stimuli defines the locational types that are available or not available, chosen or rejected by the movers. Once again, the multidimensional scaling algorithm is the major analytic tool. The results suggest that revealed preference analysis could be used to uncover the preference of urban movers with a high degree of reliability.

The paper by Burnett examines data problems in the application of conjoint space analysis to the selection of urban travel modes. The recurrent urban travel problem is defined in the context of intraurban shopping in Dallas, Texas. This type of study obviously involves data collected in the field rather than in the laboratory and provides an interesting point of comparison with the Monte Carlo simulation studies discussed in Part 1. Specific problems investigated are concerned with the complexity of data-gathering procedures, the independence of attribute assumption, the additivity assumption, and computational and interpretive procedures. The author claims that the usefulness of the conjoint model in field experiments is exemplified in the solutions that she achieved to each of these problems.

The paper by Louviere provides a set of theorems that describes mathematical procedures for establishing empirical laws in spatial decision making. His aim is to derive a set of theorems that permits one to test models based on entirely physical (or observable) values. However, he does go on to show that, where physical values are not available, measurements can be inferred once the empirical laws related to the physical values are established. In short, the author contends that the establishment of meaningful empirical laws is the first-order requirement for the study of spatial phenomena and focuses attention on the problems that data analysts are exposed to when they assume a single model and transform the data to fit such a model. His paper also is illustrated with a number of examples.

A variety of problems faced by researchers in geography, marketing, and psychology are presented in Part 3. Pipkin begins by summarizing some of the major shortcomings of current work in the interface between geography and psychology. Both the examples of his own work quoted in his paper and the next four contributions emphasize the methodological, conceptual, epistemological, and practical problems of using complex data sets. For example, Golledge, Rayner, and Rivizzigno focus on problems associated with collecting a satisfactory set of data, analyzing it by

MDS methods, and interpreting the results. Their paper focuses on a multidimensional analysis of cognitive representations. It discusses an incomplete experimental design for the collection of data in a questionnaire-type format and then concentrates on the problem of comparing output from a multidimensional scaling algorithm with an actual configuration of data points. The techniques used in the comparative stage are two-dimensional spatial correlations and two-dimensional spectral analysis.

Though many of the preceding papers use multidimensional scaling in one way or another, the paper by Bernard Marchand aptly highlights two serious problems that have to be considered by anyone undertaking a multidimensional scaling analysis of large data sets. The first problem is that of identifying dimensionalities of the spaces that are recovered. The second problem is one that is rarely, if ever, taken into consideration either in analyzing configurations from a multidimensional scaling algorithm or, indeed, in data analysis generally. This is the problem of fractional dimensions of space. Marchand's paper is brief, to the point, and problem oriented; and it leaves us with the very clear opinion that, while considerable work using multidimensional analytical techniques has been completed in recent years, there are still significant and unanswered questions that are extremely relevant to the multidimensional analysis of data sets in general.

The question of reliability in data sets is discussed by Deutscher, who summarizes the range of multidimensional scaling studies used within a marketing context and points out that the dominant type of data used in these studies is paired similarity judgments made by consumers. His paper includes a review of the marketing literature on data collection and comments critically on the quality of such data. A measure of the reliability of paired data comparisons is described, and the output from using this measure is presented. Of particular significance are comments concerning the consistency and reliability of data over time and the usefulness of reliability measures in defining spurious shifts in subjects' statements over time.

Continuing with the theme of reliability, Green, in the final paper in this volume, defines three components of reliability that he uses as an aid in locating sources of measurement error, particularly in scaling analysis. These components are *the congruity of the data, the fit of the model,* and *its generalized ability.* He utilizes a sampling of observations from large sets of data to define correlation indexes for assessing each component. The examples used are drawn from a Monte Carlo analysis of interactive scaling [6]. He examines in detail the source of excessive measurement error by inspecting the level of each index and relations among the indexes. The final step is describing a procedure for testing generalized ability relative to other indexes.

The papers presented in this volume represent some of the interactions occurring along the interface between geography and psychology. As pointed out in the Introduction, these interactions have been going on for some time and are more widespread than those represented in this sample. Despite considerable variation in the topics covered, each of the papers in this volume is directed toward the solution of a set of serious problems relating to experimental design, data reliability, and the separation of analytical methods that are appropriate for data sets of various sizes, completeness, and reliability. It is hoped that this volume will encourage many more interactions among geographers, psychologists, and other social scientists interested in theoretical and practical problems of data and measurement and the innovative use of multidimensional methods of analysis.

Reginald G. Golledge

John N. Rayner

REFERENCES

1. Coombs, C. H., 1964. *A Theory of Data*. New York: John Wiley and Sons.
2. Golledge, R. G., and G. Rushton, 1972. *Multidimensional Scaling: Review and Geographic Applications*. Washington, D.C.: Association of American Geographers, Commission on College Geography, Technical Paper #10.
3. Golledge, R. G., and G. Rushton, (eds.), 1976. *Spatial Choice and Spatial Behavior*. Columbus, Ohio: Ohio State University Press.
4. Green, R. S., J. T. Kehoe, and T. J. Reynolds, 1975. "A Monte Carlo Evaluation of Computer Interactive Ordering for Individual Subjects," a paper presented to Psychometric Society Meeting, Iowa City, Iowa, April 1975.
5. Isaac, P., and D. Poor, 1974. "On the Determination of Appropriate Dimensionality in Data with Error," *Psychometrika*, 39, 91-109.
6. Klahr, D., 1969. "A Monte Carlo Investigation of the Statistical Significance of Kruskal's Nonmetric Scaling Procedure," *Psychometrika*, 34, 310-30.
7. Shepard, R. N., A. K. Romney, and S. B. Nerlove, 1972. *Multidimensional Scaling: Theory and Applications in the Behavioral Sciences* (2 vols.). New York: Seminar Press.
8. Spence, I., 1972. "A Monte Carlo Evaluation of Three Nonmetric Multidimensional Scaling Algorithms," *Psychometrika*, 37, 461-86.

9. Spence, I., 1974. "On Random Rankings Studies in Non-Metric Scaling," *Psychometrika*, 39, 267-68.
10. Spence, I., and N. W. Domoney, 1974. "Single Subject Incomplete Designs for Non-metric Multidimensional Scaling," *Psychometrika*, 39, 469-90.
11. Wish, M., 1972. "Notes on the Variety, Appropriateness, and Choice of Proximity Measures," a paper for the Workshop on MDS, June 7-10, 1972, Bell Laboratories, Murray Hill, N. J.

Introduction — Substantive and Methodological Aspects of the Interface between Geography and Psychology

Reginald G. Golledge

The last two decades have seen an increase in the number of cross-disciplinary interactions in the social and behavioral sciences. This is part of the breakdown of strict boundaries between the disciplines and reflects a desire to gain an increased understanding of both human and physical phenomena and the relationships between them. Although there have been ongoing casual flirtations between geography and psychology for the better part of this century — particularly with respect to the use of maps and the understanding of space and spatial orientation — there was, prior to 1960, no clearly identified strong link between the two subjects in the way that a clearly identified link existed between geography and economics, geology, meteorology, anthropology, and sociology.

The reasons for this lack of interface are obvious. For many years geographers were concerned only with macroenvironments (in a psychological sense) and cared little about fundamental psychological variables such as learning, preference, choice and decision making, attitudes, images, values, personality, motivation, and so on. Geography was dominated by a form-oriented approach, which took as its fundamental data sets the environment external to humans and the modifications made to the external environment by humans working alone or in groups. In other words, geographers were interested in the spatial manifestations of human existence and searched for explanations in terms of coincidental relationships between overt human activity and structural characteristics of the environment and its resource base. With few major exceptions, the field of psychology concentrated on the microscale, with an emphasis being placed on the processes by which sentient beings were able to manifest behavior. Critical variables often were defined on humans or other thinking organisms in experimental situations. Broad relations among individuals and manifestations of mass behavior were seen to be more the prerogative of sociologists.

The development of process-oriented approaches for analyzing various aspects of human geography inevitably led to the discipline of psychology. This was true when description of "man/environment" relations gave way to more extensive searches for explanations of such relations. In addition to describing location patterns and attempting to define optimal spatial arrangements of such patterns, there emerged a concern for knowing why particular patterns exist; attempts were made to break down the human decision-making process into its component parts so as to find which parts had a pronounced and fundamental spatial component. At the same time, in the discipline of psychology, there emerged a growing awareness of the social aspects of human existence and an even more fundamental realization that humans exist in complex external environments that have an impact on the way they live and act.

In one of the earlier publications linking geography and psychology [35], Lowenthal argued that the universe of geographical study can be divided into three broad realms: the first being the nature of the environment; the second, what humans think and feel about the environment; and the third, how humans behave in the environment and alter it. While stressing that these realms are interrelated and cannot be understood in isolation, he suggested that they require different analytical techniques and are usually studied by separate disciplines. His point was that until the early 1960s geographers were content to explore only the first realm (i.e., the real world). He further pointed out that in daily practice "objective reality" is subordinated to the world that is perceived, experienced, and acted in. His argument suggested that it is the processes by which human obtain their impressions, develop their values, experience, and act that lay at the heart of the developing interface between geography and psychology.

SOME COMMON CONCERNS
AND COMMON PROBLEMS

Geography is a science that concentrates on the location and distribution of phenomena, the interactions among them, and the processes that produce those locations, distributions, and interactions. It is, therefore, as much concerned with the world outside of a behaving organism as it is with the elements of the behaving organism itself. Whereas the psychologist may be interested in the physiological activities of the organism (including such things as the way senses perceive or are stimulated, the physiological structure of the organism and its interrelated parts, its maturation and developmental processes, its feelings and mental health, and so on), many of these activities may be of little direct concern to the geographer. For the most part, geographers have been interested in overt behavior, and only recently have they become more conscious of the human behavioral processes that produce overt behavior. They are still only partly interested in covert behavior and have shown little or no interest in the physiological basis of behavior, except for some passing interest in the idea that knowledge of spatial concepts may be highly localized in the brain.

For the psychologist, behavior includes those activities of a sensate organism that can be observed by other organisms or by an experimenter's instruments. Observations of behavior may be direct and unaided or indirect and instrument aided. Although there may be some disagreement concerning the relative emphasis placed upon "lower organisms" as compared to humans, there is little doubt that both types of organisms are valid subjects for psychological study. Psychologists may be interested in the physiological activities of grazing cattle, but geographers never have been. Geographers rarely consider "lower organisms" in any sense other than in trying to understand how humans use them or control their distribution for their use.

A common concern between geographer and psychologist is to discover regularities in the relationships among variables. Such relationships provide a basis for understanding patterns that may be repetitive, whether in a behavioral sequence or in a spatial distributional sense. A long-standing area of psychology is experimental psychology, which is grounded in laboratory method. But, for most of geography's existence, only the physical side of geography has extensively used the laboratory method. During recent years, however, some human geographers have begun to experiment in laboratory-type situations and to develop a vocabulary of experimental and analytical methods that are easily understood and responded to by psychologists.

The fact that there is continual change in human experience and behavior has been recognized throughout the history of humankind. During this century, change may have progressed faster than during any previous time, and it appears to be accelerating. Much of this change is clearly identifiable in political, social, and economic terms. There are also substantial changes taking place in the natural environment as resources are depleted and developed and as humans construct a new environment to house their ongoing activities. Although psychology deals with very few, if any, of these large-scale developments, it does deal with the raw material out of which they are shaped, namely individual human beings. Fundamentally, one can take the position that, in order to understand this process of continual change in human experience, one has to understand the individual and relations among individuals. Although geography traditionally has dealt with *manifestations* of this process of change in both the natural and built environments, it has only recently accepted the idea that an understanding of the nature of such changes necessitates some knowledge of individuals and their thoughts and actions.

Though most of the early borrowings went from psychology to geography, there has been a growing interest within psychology in the study of the human's relationship to the everyday physical environment. While geographers began thinking about the world in the human mind, psychologists became increasingly aware of the position of humans in the world at large.

One of the most fundamental problems facing researchers interested in "man-environment relations" was one of communication. For example, an examination of the geographic literature shows that the borrowing of terms from psychology was often accompanied by a less than accurate interpretation of the meaning of the terms. The use of the term *perception* is a classic example of this. Many geographers used the term *perception* as a synonym for *cognition*, but they did so without an understanding of the breadth of the concept of cognition or the narrowness of the concept of perception. Similarly, many geographers confused the concepts of *behaviorism* and *behavioralism*.

Problems of communication were partly overcome by communal reading of basic theory (including choice and decision theory, data theory, and learning theory, particularly the various works of Piaget) and by a common concern with methodology. The increasing concern of geographers—who had just passed through a "quantitative revolution"—with precision, measurement, model building, prediction, and improved explanation merged with the aims of several other disciplines, foremost of which were economics and psychology. Where verbal communication broke down, a common statistical and mathematical base helped renew bridges between the disciplines. At this stage, most of a geographer's statistical training was obtained from courses offered in psychology, educational psychology, or mathematical economics and econometrics. Minimal mathematical training (including calculus and linear algebra) also improved the ability of members in each discipline to communicate and to comprehend each other's academic research.

Despite the communality induced by analytical methods, the steadily accumulating body of information on topics of common concern (e.g., humans and their relation to the environment) continued to produce problems because of the variety of languages, techniques, and models that were employed in research. Such differences often obscured many similarities in the nature of the problems being considered. One had to read diligently in order to understand the relationship between economic aspects of environmental quality and psychological investigations of the same problem; to some extent, both disciplines found a bridge in the geographer's attempts to model and explain similar problems.

HISTORY OF THE SUBSTANTIVE INTERFACE
BETWEEN GEOGRAPHY AND PSYCHOLOGY

If one searches diligently enough, one can almost invariably find evidence of overlap between any selected pair of scientific disciplines. It is not my purpose in this paper to search diligently to find coincidental areas of overlap between geography and psychology. Although it is appropriate to make reference to a few key works that are traditionally recognized as being important connectors between the two disciplines, I would like to concentrate on the period from 1960 onward, for it is during this period that the most consistent and conscientious cross-disciplinary activities have take place.

Some early evidence of the relationship between geography and psychology in a substantive sense appears in the work of C. C.

Trowbridge [47], an American scientist who investigated people's methods of orientation and imaginary maps. Interestingly enough, this initial systematic experiment on human knowledge of large-scale environments seems to have been markedly influenced by the work of a geographer, F. P. Gulliver (1908) [25], who found that the school-age children he studied oriented themselves in the environment more by egocentric orientation than by cardinal compass directions.

During the decades following Trowbridge's work, geography was dominated by environmental determinism, which tied together psychological and physiological aspects of humans with a variety of physical characteristics of the large-scale environments in which they lived. As a part of the reaction against environmental determinism, geographers moved more toward examining the form and content of environments and the patterns of human activity over the surface of the earth in a relatively objective manner. In doing so, they generally divorced themselves from the link that had begun to develop with physiological psychology, personality theory, and behaviorism.

From the 1920s to the late 1950s, geographers generally maintained their interest in macroscale problems. Meanwhile, psychologists became more experimentally inclined, and there was greater emphasis on laboratory experimentation and abstract model building. It was not until there was a revolution in geographic thought and methodology that the potential for the two disciplines to interface once again became apparent. Initial contact was made via reference to a number of important books and papers, such as those by Firey [13], J. K. Wright [55], Tolman [46] and later those by Boulding [4], Lynch [39], and Strauss [45].

From 1960 on, the number of cross-disciplinary contacts increased dramatically. Two of the earliest points of contact (which have had the greatest impact in the cross-fertilization of ideas between the two disciplines) were in the areas of decision making and choice theory on the one hand and perceptions of, and attitudes toward, natural environments on the other hand. During the 1950s, geographers had been stimulated by a series of innovative papers on the diffusion of innovations written by Torsten Hägerstrand, who had grasped the potential of using Monte Carlo simulation studies for examining various spatial problems [26, 27, 28]. Although at this time geography leaned much more heavily on economics, anthropology, and sociology than it did on other disciplines, there was evidence that an interest in some aspects of microeconomics (e.g., decision theory and choice theory) could be complemented by a parallel interest in psychology on decision making and the choice act in order to obtain a more complete understanding of the reasons for human activities in space.

During the 1960s, therefore, geographers began exhibiting a general interest in subjective aspects of individual behavior, and a school of thought emphasizing a process approach to the study of the relation between humans and their environment (called the behavioral approach) emerged. This movement paralleled the more dominant and more highly articulated subsets of the discipline that concentrated on the structure of human activities in the environment and on the environment itself. Critical areas of geographic research that began using theories and methods from psychology included the research program at the University of Chicago on the perception of natural hazards (under the leadership of Gilbert White and the able direction of Robert Kates and Ian Burton); the more humanistic work of David Lowenthal and others on the perception of landscapes; and a variety of approaches to decision making, search behavior, spatial learning, and spatial choice by Julian Wolpert [53, 54], Peter Gould [23, 24], and Reginald Golledge [14].

In the area of environmental perception, White and his colleagues investigated the relation between peoples' beliefs about natural hazards and their subsequent behaviors. Research into the somewhat unusual pattern of behavior of continuing to live in hazardous areas indicated quite clearly that many individuals were dominated by short-run goals and that their behavior was suboptimal.

In a similar way, Wolpert investigated the pattern of farming activities in Sweden and indicated quite clearly that economic rationality was by no means a dominant force behind the decisions of farmers to select crop types [53]. By introducing Simon's notion of "satisfaction," rather than "economic rationality," as a criterion for evaluating human actions and by using the concept of boundedly rational spatial behavior (adapted from Simon's classic works), Wolpert provided a clear alternative to existing economics-biased geographic theory and paved the way for the reintroduction of studies of individual and small-group behavior into geography. This was seen as a clear alternative to the large-scale studies that had begun to proliferate in the discipline.

Gould, in 1963, indicated quite clearly that the pattern of human activities in the environment could be interpreted differently, depending on one's attitude toward risk [23]. His work on game theory was seminal in geography. In developing the notions that humans at the individual and group level develop various strategies to aid in their exploitation of nature's resources and that varying patterns of activity could be best explained by uncovering the strategies used in choice- and decision-making processes, Gould paved the way for an extremely strong tie with important areas of psychology.

In their work on attitudes toward different environments, Lowenthal [34, 35] and Lowenthal and Prince [38] showed that landscapes themselves reflect the attitudes of the humans who developed and modified them and that landscapes can best be understood when the attitudes, preferences, values, and beliefs of individuals are understood. This line of research continued as geographers (such as Tuan [48]) examined changing conceptions of the environment as seen over historical time (geosophy).

Though there were other streams that emphasized the changing substantive area of human geography (e.g., ethnogeography and diffusion theory), perhaps the most recent strong link forged between geography and psychology came through a discussion of space-learning activities. This line of research, on the one hand, concentrated on mathematical learning models and their possible application to spatial behavior [14, 16, 17] and, on the other hand, on development and discussion of the concept of cognitive mapping and cognitive images [8, 9, 10]. Emerging also at this time were two other areas of cross-disciplinary contact, including the work of Gould on mental maps and preferences for different environments, the work of Rushton on revealed preferences for places, and a series of papers by David Huff on consumer behavior.

At this time, therefore, a number of different and rather disparate strands in geography all developed ties with different subsets of psychology. Hazard researchers developed links with the fields of perception and attitude formation and used measurement techniques such as semantic differential scales and other unidimensional scaling procedures. Researchers in choice and decision theory were concerned with substantive areas in psychology and economics and adapted game theory and decision theory models for use in geographic situations. Consumer behavior studies drew partly on cognitive psychology and partly on learning theory and methodologically introduced mathematical learning models, Luce's choice theory, and methods such as multidimensional scaling and repertory grid analysis to geography. In addition to these links, a wide variety of human-oriented studies found relevant theory in psychology—such as personality theory, developmental-constructionalist theory, perceptual theory, learning theory, motivation theory, and so on.

On of the immediate results of this reaching out for knowledge by geographers was the happy coincidence of finding both theories and models suited for explaining the type of empirical situations in which they were interested. In consumer behavior, it became obvious that economic rationality did not dominate consumer's actions, unless one could combine the abstract concept of economic rationality with theories of perception and cognition and assume that consumers acted *as though* they were economically rational *given their information base*. In hazard research, it was found that the actions of individuals who remained in hazardous areas even where the probability of future exposure to hazards was high could be partly explained by cognitive dissonance theory and partly by information on habits, attitudes, and the process of forgetting.

Since that burst of development during the 1960s, cognitive mapping, environmental perception, environmental cognition, analysis of preferences, choice theory, risky decision making, spatial learning, spatial imagery, environmental personality, and artificial intelligence have all been substantive areas of overlap between the two disciplines. Perhaps the most marked, however, has been the broad area of environmental cognition and mapping, where the interface between geography and psychology has been complemented by cross-disciplinary research with other areas such as planning, architecture, sociology, and anthropology. Representatives of each of these disciplines join together at annual Environmental Design and Research Association conferences to further enhance the cross-fertilization of ideas and methods.

The need to build bridges between disciplines interested in environmental problems was stated clearly in a special issue of the *Journal of Social Issues* (October 1966 [32]), edited by geographer R. W. Kates and psychologist J. F. Wohlwill. In his essay [31], Kates suggested that the social and behavioral sciences could help bridge the gap between all environmental disciplines by relating science and design to human needs. He went on to argue that "in building such a bridge, geography and psychology, in particular, might find the beginning of modest but worthy collaboration" [31, p. 26]. He pointed out that, though geography had long dealt with the stimulus properties of the environment, psychology had mainly ignored the large-scale external physical environment, even though it may have intensively studied perception and symbolization of discrete stimuli in the environment. His suggestion that geographers consider the theories and practices of psychology in their studies of environment and that psychologists move beyond the laboratory into the problems of the real world itself could serve as two endpoints of a bridge between the disciplines. His reference to the increasing number of psychologists who were dealing with the psychology of space (apart from the extensive literature on perception) indicated those strands on which a bridge could be built.

In that same issue, Wohlwill [52] argued that, though psychologists never tire of pointing out the importance of stimulus factors as determinants of behavior and the role of the environment in influencing behavior, they had spent little time or effort on examining problems related to the human response to the physical environment. Wohlwill pointed out how child psychologists, personality psychologists, and social psychologists all had environmental biases in their work and how they should complement their biases with more extensive knowledge of physical environments. Quoting Hebb's influential work [29] that focused on sensory stimulation from the environment and its relevance for normal development of perceptual and cognitive functions as well as motivational processes, Wohlwill argued that psychologists should go beyond their experimentations with animals and think more in terms of the relationship between humans and environment itself. Further reference to work by Calhoun (on animals and their physical environments) and Sommer (on human personal spaces) indicated that many links between geography and psychology already existed, though they had not been widely publicized.

Meanwhile, geographers such as Lowenthal and Sonnenfeld, following the suggestions of J. K. Wright, expressed interest in the concepts of environments being imagable and having personalities. Environments were also seen as generating potential stress situations, and the concept of stress was shown to have widespread application to movement behavior by Wolpert [54], who cast migration in the role of a response to environmental stress and to consequent changes in place utility. The emerging cross-disciplinary environmental concern movement was also expressed in works such as that by Lucas on wilderness-policy decision making and recreational environment.

Although the sources of the links between geography and

psychology were many, there were also some very obvious differences in the manner in which links were forged. Both geography and psychology are broad disciplines, and the relative significance of theory, model, and measurement varies considerably across different parts of each discipline. Some of the earlier links between geography and psychology were conceptual and there was little follow-up of measurement procedures, but other ties were more strongly oriented toward measurement theory and model building. A good part of behavioral research in geography, particularly that articulated in journals such as *Environment and Behavior* and *Geographical Analysis*, was rigorously statistical in nature and model building in orientation and appeared closely tied to decision making, choice, and learning theories, and cognitive representations of space. Work in other areas was more solidly linked to concepts of values, morals, ethics, images, and subjective methods of evaluation and appeared more in the *Annals of the Association of American Geographers* and the *Geographical Review.* In particular, the environmental perception literature became in part "metaphysical in outlook, anthropological in content, and historical and literary in their modes of analysis and presentation" [36, p. 252]. Although both lines of thought strengthened the interface between the two disciplines by suggesting new modes of inquiry based on holistic concepts of human and milieu, there was a very clear distinction—which focused on the emphasis given to data theory, measurement theory, and model building—between the two types of linkages. As an example of the measurement- and model-oriented approach, a 1969 volume called *Behavioral Problems in Geography* [7] emphasized theoretical and model-oriented links between geography and psychology. Unlike many of the papers in the environmental perception literature, this volume included topics such as inference problems in locational analysis, conceptual and measurement problems in cognitive behavioral approaches to location theory, interdependencies in locational decisions, friends and neighbors voting models, the geographical relevance of learning theories, the spatial structure of acquaintance fields, place utility and intraurban migration, the scaling of locational preferences, the measurement of mental maps, and the modeling of interactions. The contributors were geographers, a regional scientist/planner, and an architect/psychologist. The papers were united by a common concern for the building of future geographic theory on the basis of postulates regarding human behavior and were further united by a common concern for molding psychological and geographical theory, measurement techniques, and models. The emphasis in this book was more on understanding the processes that produced spatial manifestations than on examining the spatial manifestations themselves. One immediate result was an extended series of discussions in the discipline of geography on the relative importance of form- and structure-oriented approaches as opposed to process-oriented approaches.

A special edition of *Environment and Behavior* in 1972 epitomized the work on environment and behavior in terms of personality types, epistemological modes of apprehension, varieties of individual and community spaces, and connections between environmental and social interactions. Contributors to this volume included Craik (a psychologist) and Lowenthal, Sonnenfeld, Buttimer, and Tuan (all geographers). An earlier monograph edited by Lowenthal had included contributions by geographers, a psychologist, and planners [35]. Topics covered included attitudes toward environment, spatial meaning, and the properties of the environment, environmental perception and adaptation levels, perception of storm hazards, and the "view from the road."

Much of the early interface between psychology and geography was concentrated in a limited number of universities, including Clark, Pennsylvania State, Iowa, Ohio State, and Michigan. Substantive points of contact included hazard perception, environmental learning, space perception, decision theory, choice theory, search processes, environmental knowing, and space preferences. Methodologically, interest was focused on attitude scales, repertory grid techniques, Luce's choice theory, mathematical learning models, multidimensional scaling, cognitive cartography, individual differences scaling, and various aspects of experimental design.

In psychology, perhaps some of the earliest conscious interactions with geography and geographers were those undertaken at Clark University, particularly in the area of perception and environmental learning, when researchers such as Wapner, Beck, and Stea found an interest expressed by geographers in using developmental theories from psychology as an aid to their understanding of space and spatial concepts. This growing interest in the environmental area became centralized at the City University of New York, where the innovative work done by Proshansky, Ittelson, and Rivlin was supplemented by the emergence of a new journal, *Environment and Behavior*, edited by Gary Winkel, who, by that stage, had extensive contacts with geographers in both

Canada and the United States and who was frequently invited to criticize and present papers at their meetings.

For many geographers, the influence of Clyde Coombs and his group of mathematical psychologists at the University of Michigan was quite noticeable. In other universities, there was a growing concern for reading the literature of the other discipline, but the degree of interpersonal contact and interaction perhaps nowhere else reached the degree of cooperation and joint work that it did at Pennsylvania State, Iowa, and Ohio State.

The 1970s saw a consolidation of the link between geography and psychology and an increasing interawareness. This awareness has been fostered by the production of books with titles such as *Image and Environment* [10], *Maps in Minds* [11], *Environmental Knowing* [40], *Environment and Behavior* [41], *Environmental Psychology* [42], *An Introduction to Environmental Psychology* [30], *Environmental Planning* [44], *Cities, Space and Behavior* [33], and *Spatial Choice and Spatial Behavior* [20]. Within both disciplines there has been a trend toward incorporating models, concepts, theories, and measurement practices from the other discipline into textbooks as well as into teaching and research operations.

The potential of each discipline for contributing to the other continues to be realized. Some basic psychological theory is now taught as a part of many introductory geography classes; geographic concepts of space and organization and the analysis of pattern and distribution are creeping into some areas of psychology. The substantive links have been forged, and a variety of these links are illustrated in the papers in this volume.

COMMON CONCERNS WITH METHODS AND DATA

There has also been considerable borrowing and adaptation of methods and techniques between geography and psychology. In the substantive area, it was primarily the geographers who made excursions into psychology and who developed the first lines of contact between the disciplines. Just as in the substantive area, much of the early methodological contact between geography and psychology was in the form of imitation and borrowing of methods from psychology by geographers. During the 1950s, correlation and regression analysis dominated human geography, and, though some of their models came from economics, in many cases the methods were introduced by geographers who had been exposed to statistics courses in psychology departments. This trend continued through the 1960s and is still a common one today, although most geographers now get their statistical training within their own discipline.

Apart from correlation and regression analysis, perhaps the most important borrowing by geography from psychology was related to grouping techniques such as factor analysis and discriminant analysis. For almost a decade, factorial ecology (factor analysis of the socioeconomic and cultural groupings of people in cities) dominated segments of the discipline. Since that time, however, there has been an increasing interest among geographers in psychometric methods such as uni- and multidimensional scaling, clustering, analysis of individual differences, repertory grid analysis, conjoint measurement procedures, and, of course, a wide range of nonparametric statistical procedures.

Borrowing by psychology from geography has been more a trend of the 1970s than of the 1960s. In the general area of environmental cognition, the geographer's favorite analytic device—the map—has come to be extensively used in psychology. Along with mapping techniques, a range of different spatial statistics oriented toward developing a capability for measuring characteristics such as proximity, density, and dispersion (e.g., in crowding research), nearest neighbor analysis, and both temporal and spatial seriation methods (e.g., spectral analysis and spatial autocorrelation) have become more widespread in psychology. Another favorite analytic tool of the geographer, the location-theoretic model (when developed to the location-allocation model) was found to have a direct appeal for many data analysts and methodologists in psychology.

While geographers were continuing their search of the psychology literature for substantive and methodological areas of interest, some psychologists were moving out of the laboratory and into the macroenvironment in their search for an understanding of individual behavior. This movement was formalized during the 1970s under the heading of "environmental psychology" and has continued to be a major area of interface between the two disciplines. In particular, the book of readings on *Environmental Psychology* [42] proved to be a significant contact point between the disciplines, as did the development of the Environmental Design and Research Association.

Traditionally, geographers have been interested in populations of normal adults. During the 1960s, they developed an interest in many of the subject populations that typified work in psychology, including populations of children, students, minority groups, and various disadvantaged groups such as the aged, the infirm, the mentally handicapped, and the poor. In terms of subject population, geographers also have dealt primarily with large groups of subjects—including entire nations—and have used agglomerations of data such as might be found in census volumes. The use of small subject sets (five to ten subjects or fewer) was frequently criticized. The deeper probing done by psychologists frequently demanded these small sets, and rarely did they concern themselves with populations of a magnitude similar to that traditionally analyzed by geographers.

As members of each discipline became more aware of the advantages and disadvantages of the other discipline in accessing and dealing with subjects, a trend emerged for psychologists to increase the size of their population subgroups and for geographers to decrease theirs. In many cases, geographers retreated to small samples and to experimental or laboratory situations, exactly the situations that psychologists were breaking out of. It almost looked as though the geographers in their haste to become more systematic and thorough and the psychologists in their haste to become more aware of the complexities of large-scale environments would pass by each other as they proceeded vigorously in their opposite directions. Although this happened in a few cases, opportunities for discussing the advantages and disadvantages of different subject types and subject size groups abounded. Both disciplines are now aware of the advantages and disadvantages of dealing with both large and small data sets, but there is still the tradition of using large data sets and macroscale analysis in geography and small data sets and microscale analysis in psychology.

The desire to exchange information on problems of handling data stimulated the development of the conference from which many of the papers in this book originated. It was obvious that each discipline could pass on information of considerable importance to the other in terms of understanding and dealing with the complexities of handling data sets with different orders of magnitude, different degrees of reliability, and different levels of completeness. The results of the conference and subsequent work are, of course, presented in this volume.

For the most part, the geographer's data have been objectively collected by disinterested observers. Much of the data relate to what psychologists would call the "antecedent conditions" of populations—including such things as locational characteristics, or general situational variables, and structural and functional characteristics such as sex, number, age, income, education, ethnic affiliation, cultural groupings, and other miscellaneous material that could be easily collected on a census form. The psychologist, on the other hand, still relies largely on observational data collected at the individual level. Psychological data include statements of preference or judgment; sortings of categories; measurements of proximity, dispersion, degree of cluster; interval and magnitude estimates of quantities; ideals, attitudes, values, and beliefs; and so on. Both the geographer and the psychologist are interested in patterns, regionalizations or clusterings, and orderings. The potential for interaction between the two disciplines in terms of data, therefore, is obvious, even though the disciplines work on substantially different scales.

One event contributing to the methodological interface between geographers and psychologists was the seminar offered at Bell Labs in 1972 on scaling, clustering, and multidimensional analysis. At that conference, psychologists, geographers, and members of other disciplines gathered to hear updated and innovative explanations of techniques such as multidimensional scaling, individual differences scaling, three-mode factor analysis, and a range of multivariate analytic methods.

Following the Bell Labs conference, the dissemination of computer materials in various Bell Labs packages also facilitated the development of a common language and common methods for the two disciplines. In addition, the distribution to geographers of technical reports from the Thurstone Psychometric Laboratory and the Michigan Mathematical Psychology Program and geographers' awakening interest in journals such as *Psychometrika*, *Journal of the American Statistical Association*, and the *Journal of Mathematical Psychology* provided important avenues of communication among members of the disciplines. With the publication in geography of books on topics such as spatial autocorrelation and with an increasing number of papers on location-allocation algorithms, network algorithms, shortest path algorithms, and so on (many of which were concentrated in the journal *Geographical Analysis*), the methodological similarity between the disciplines became even more evident. For that section of psychology dealing with space perception and psychological distance, for data and measurement theorists in both disciplines, and for that area of geography dealing with human behavior, distance, and spatial measurement, it appeared almost inevitable that the two disciplines would move closer and closer together.

Paralleling the methodological interests was a growing concern for two similar classes of problems. Psychologists for years had investigated various experimental designs in an attempt to make their conclusions as relevant as possible to larger-scale populations. Geographers, meanwhile, had been concerned with developing various sampling procedures and finding economies in data-collection procedures that would allow them to work at the scale they wanted. The complexities of collecting data from individuals had led psychologists to experiment with latin squares and incomplete, cyclic, and random designs for collecting data; and their methods appeared eminently suitable for geographers as they began to collect data at the individual level. Common problems such as treatment of ties, treatment of missing data, violations of triangle inequality, and asymmetry in data structures have emerged in each discipline.

As geographers passed through the quantitative revolution and increased their analytical powers, therefore, they came face to face with many of the problems that had troubled other behavioral and social scientists for decades. The first was the looseness of the definition of many concepts in geography (even fundamental concepts such as location, distribution, pattern, representation, distance, proximity, and association). Considerable time and effort has been invested in tightening those concepts, delineating more clearly their definitional structure, measuring them, introducing statements of reliability, conforming to principles of data theory, and understanding and using more complex methods of experimental design.

Since the geography/psychology interface became a reality, geographers have made widespread use of techniques such as the semantic differential technique, categorical measurement procedures, ratio scaling, stochastic decision theory, game theory, Markov modeling, linear learning models, multidimensional scaling methods, individual difference scaling, graph-theoretic measures, clustering techniques, principal components analysis, factor analysis, discriminant analysis, and a wide range of nonparametric techniques. In terms of experimental design, the geographer is using methods of experimenter observation, role playing, model building, and individual data collecting such as may be required in various uni- and multidimensional procedures.

However, the attempts to use the models, methods, and theories of one discipline to provide insight and increased understanding in the other have, in turn, produced a range of problems that neither discipline has satisfactorily solved. Geographers, accustomed to dealing with massive data sets, often found the methods developed in psychology for use with much smaller subject and stimulus sets to be inappropriate. Psychologists, who had developed complex methods for handling relatively small subject groups, suddenly found themselves faced with analyzing massive data banks. Both disciplines were aware of and used experimental designs that allowed extensive sampling procedures or incomplete data-collection procedures to exist. In many cases, sampling produced no problems, for appropriately selected statistical tests could be tied in through significance measures to the sample size and the test results. However, especially in the area of multidimensional analyses of various types, methods developed for data manipulation were initially limited to data sets that could be handled on calculators and then on small- to moderate-size computers; even now with the largest computers available there is considerable difficulty in handling data sets of the magnitude that geographers frequently deal with. For example, much of the multidimensional analytical techniques depend on matrix manipulation. In many cases, matrix-manipulation methods are tied to iterative solution procedures. For the geographer accustomed to dealing with 100 counties in a state or 5000 districts in a region, the problems of data manipulation using the entire data set became horrendous. Even for dealing with relatively small subsets of data, manipulation problems are immense.

This particular set of papers has been collected because of the growing concern in both disciplines for solutions to problems associated with the use of large data sets. The papers cover a variety of topics from questions of experimental design to the practical problems of manipulating specific types of data sets. Although no claim is made that solutions are given to the many problems currently faced by each discipline, it is hoped that this book will serve to increase the interaction potential of the two disciplines in future years.

REFERENCES

1. Beck, R., 1967. "Spatial Meaning and the Properties of the Environment." In D. Lowenthal (ed.), *Environmental Perception and Behavior.* Chicago: University of Chicago, Department of Geography Research Paper #109.

2. Blaut, J. M., 1961. "Space and Process," *Professional Geographer*, 13 (4), 1-7.
3. Blaut, J. M., and D. Stea, 1969. *Place Learning*. Worcester, Mass.: Clark University, Graduate School of Geography, Place Perception Research Report #4.
4. Boulding, K. E., 1965. *The Image*. Ann Arbor, Mich.: University of Michigan Press.
5. Briggs, R., 1972. "Cognitive Distance in Urban Space." Ph.D. dissertation, Department of Geography, Ohio State University, Columbus, Ohio.
6. Brown, L. A., 1968. *Diffusion Dynamics*. Lund: Gleerup, Lund Studies in Geography B #29.
7. Cox, K. R., and R. G. Golledge (eds.), 1969. "Behavioral Problems in Geography." Evanston, Ill.: Northwestern University, Department of Geography, Studies in Geography #17, pp. 101-45.
8. Downs, R., 1967. "Approaches to and Problems in the Measurement of Geographic Space Perception." Bristol, England: University of Bristol, Department of Geography Seminar Paper, Series A #9, pp. 1-17.
9. Downs, R. 1970. "The Cognitive Structure of an Urban Shopping Center," *Environment and Behavior*, 2, 13-39.
10. Downs, R. M., and D. Stea, 1973. *Image and Environment*. Chicago: Aldine.
11. Downs, R., and D. Stea, 1977. *Maps in Minds*. New York: Harper and Row.
12. Evans, G. W., D. G. Marrero, and P. A. Butler, 1979. "Environmental Learning and Cognitive Mapping," unpublished manuscript, Program in Social Ecology, University of California, Irvine.
13. Firey, W., 1945. "Sentiment and Symbolism as Ecological Varibles," *American Sociological Review*, 10, 140-48.
14. Golledge, R. G., 1967. "Conceptualizing the Market Decision Process," *Journal of Regional Science*, 7 (2) (supplement), 239-58.
15. Golledge, R. G., 1976. "Methods and Methodological Problems in Environmental Cognition." In G. Moore and R. G. Golledge (eds.) *Environmental Knowing*, pp. 300-13. Stroudsburg, Pa.: Dowden, Hutchinson and Ross.
16. Golledge, R. G., R. Briggs, and D. Demko, 1969. "The Configuration of Distances in Intra-Urban Space," *Proceedings of the Association of American Geographers*, 160-65.
17. Golledge, R. G., and L. A. Brown, 1967. "Search, Learning and the Market Decision Process," *Geografiska Annaler*, 49B, 116-24.
18. Golledge, R. G., V. Rivizzigno, and A. Spector, 1975. "Learning about an Urban Environment: Analysis by Multidimensional Scaling." In R. G. Golledge (ed.), *On Determining Cognitive Configurations of a City*, pp. 109-34. Columbus, Ohio: Department of Geography and Ohio State University Research Foundation.
19. Golledge, R. G., and G. Rushton, 1972. *Multidimensional Scaling: Review and Geographical Applications*. Washington, D.C.: Association of American Geographers, Commision on College Geography, Technical Paper #10.
20. Golledge, R. G., and G. Rushton, (eds.), 1976. *Spatial Choice and Spatial Behavior*. Columbus, Ohio: Ohio State University Press.
21. Golledge, R. G., and G. Rushton, and W. A. V. Clark, 1966. "Some Spatial Characteristics of Iowa's Dispersed Farm Population and Their Implications for the Grouping of Central Place Functions," *Economic Geography*, 42, 261-72.
22. Golledge, R. G., and A. Spector, 1978. "Comprehending the Urban Environment: Theory and Practice," *Geographical Analysis*, October, 403-26.
23. Gould, P., 1963. "Man against His Environment: A Game Theoretic Framework," *Annals of the Association of American Geographers*, 53, 290-97.
24. Gould, P., 1967. "Structuring Information on Spatio-Temporal Preferences," *Journal of Regional Science*, 7, 2-16.
25. Gulliver, F. P., 1908. "Orientation of Maps," *Journal of Geography* 7, 55-58.

26. Hägerstrand, T., 1952. *The Propagation of Innovation Waves*. Lund: Gleerup, Lund Studies in Geography Series B #4.
27. Hägerstrand, T., 1965. "A Monte Carlo Approach to Diffusion," *European Journal of Sociology*, 6, 43-67.
28. Hägerstrand, T., 1968. *Diffusion of Innovations* (Alan Pred, trans.). Chicago: University of Chicago Press.
29. Hebb, D. O., 1949. *The Organization of Behavior*. New York: John Wiley and Sons.
30. Ittelson, W. H., H. M. Proshansky, L. G. Rivlin, and G. Winkel, 1974. *An Introduction to Environmental Psychology*. New York: Holt, Rinehart and Winston.
31. Kates, R. W., 1966. "Stimulus and Symbol: The View from the Bridge," *Journal of Social Issues*, 22 (4), 21-28.
32. Kates, R. W., and J. F. Wohlwill, 1966. "Man's Response to the Physical Environment," *Journal of Social Issues*, 22 (4), 15-20.
33. King, L. J., and R. G. Golledge, 1978. *Cities, Space and Behavior*. Englewood Cliffs, N.J.: Prentice-Hall.
34. Lowenthal, D., 1961. "Geography, Experience, and Imagination: Towards a Geographical Epistemology," *Annals of the Association of American Geographers*, 51, 241-60.
35. Lowenthal, D. (ed.), 1967. *Environmental Perception and Behavior*. Chicago: University of Chicago, Department of Geography Research Paper #109.
36. Lowenthal, D., 1972. "Editor's Introduction" (to a special issue on the Human Dimensions of Environmental Behavior), *Environment and Behavior*, 43, 251-54.
37. Lowenthal, D., 1972. "Research in Environmental Perception and Behavior: Perspectives on Current Problems," *Environment and Behavior*, 43, 333-42.
38. Lowenthal, D., and H. Prince, 1965. "English Landscape Tastes," *Geographical Review*, 55, 186-222.
39. Lynch, K., 1960. *The Image of the City*. Cambridge, Mass.: M.I.T. Press.
40. Moore, G., and R. G. Golledge, 1976. *Environmental Knowing*. Stroudsburg, Pa.: Dowden, Hutchinson and Ross.
41. Porteous, J., 1977. *Environment and Behavior*. Reading, Mass.: Addison-Wesley.
42. Proshansky, H. M., W. H. Ittelson, and L. G. Rivlin, 1970. *Environmental Psychology: Man and His Physical Setting*. New York: Holt, Rinehart and Winston.
43. Rushton, G., 1979. *Optimal Location of Facilities*. Wentworth, N. H.: Compress.
44. Saarinen, T., 1976. *Environmental Planning: Perception and Behavior*. Boston: Houghton, Mifflin and Company.
45. Strauss, A., 1961. *Images of the American City*. New York: Free Press.
46. Tolman, E. C., 1948. "Cognitive Maps in Rats and Men," *Psychological Review*, 55, 189-208.
47. Trowbridge, C. C., 1913. "Fundamental Methods of Orientation and Imaginary Maps," *Science*, 38, 888-97.
48. Tuan, Y., 1967. "Attitudes towards Environment: Themes and Approaches." In D. Lowenthal (ed.), *Environmental Perception and Behavior*. Chicago: University of Chicago, Department of Geography Research Paper #109.
49. Tuan, Y., 1977. *Space and Place*. Minneapolis: University of Minnesota Press.
50. Wapner, S., 1977. "Environmental Transition: A Research Paradigm Deriving from the Organismic-Developmental Systems Approach." In D. Burke et al. (eds.), *Behavior Environment Research Methods*, pp. 1-9. Madison: University of Wisconsin, Institute for Environmental Studies.
51. Winkel, G., R. Malek, and P. Thiel, 1970. "Community Response to the Design Features of Roads: A Technique for Measurement," *Highway Research Record*, 305, 133-45.

52. Wohlwill, J. F., 1966. "The Physical Environment: A Problem for a Psychology of Stimulation," *Journal of Social Issues*, 12 (4), 29-38.
53. Wolpert, J., 1964. "The Decision Process in the Spatial Context," *Annals of the Association of American Geographers*, 54, 537-58.
54. Wolpert, J., 1966. "Migration as an Adjustment to Environmental Stress," *Journal of Social Issues*, 12 (4), 92-102.
55. Wright, J. K., 1947. "Terrai Incognitae: The Place of the Imagination in Geography," *Annals of the Association of American Geographers*, 37, 1-5.
56. Zannaras, G., 1968. "An Empirical Analysis of Urban Neighborhood Perception." Master's thesis, Department of Geography, Ohio State University, Columbus, Ohio.
57. Zannaras, G., 1973. "An Analysis of Cognitive and Objective Characteristics of a City: Their Influence on Movements to the City Center." Ph.D. dissertation, Department of Geography, Ohio State University, Columbus, Ohio.

Part 1
Experimental Design
and Measurement Problems

Chapter 1.1 Surveying Multidimensional Measurement

W. R. Tobler

The largest multidimensional data set encountered in the history of geography was in the exploration of the world. One of the problems was to establish the geometric relation of all places to each other. This was to be done in a two-dimensional space of poorly known size and shape. This two-dimensional space, of course, contains an infinity of places, so some choice had to be made to determine which relations should be the first to be established. A sampling framework was required. Smaller-scale surveying is known to have been practiced in ancient Egypt, China, and Rome, but the earliest systematic efforts at exploring of the world were provided by Prince Henry the Navigator (1394-1460) of Portugal. A study of Captain Cook's voyages from the multidimensional scaling point of view should also be rewarding. It seems that the procedure was to first perform a reconnaissance, learn the general lay of the land, and then fill out the findings with details of increasing reliability and refinement. A similar process continues today, with the great continental geodetic connections occurring only in the present century.

What has all of this "old stuff" to do with the materials in the volume at hand? My contention is that many of the concepts developed during the survey of the earth are useful in the investigation of more abstract spaces. We still need to reduce the sample to manageable size and to proceed by iteration, refining each measurement as we begin to learn the structure of the relevant environment. The theory of errors plays a prominent part in modern cartography, and it has helped surveyors to formulate optimal observational plans, one of the major concerns of the papers in this volume. Thus I set the framework within which to evaluate my assigned contributions.

Four objectives of a land survey are to estimate the configuration of a set of points relative to each other, to estimate the orientation of this configuration relative to the North Pole, to estimate the absolute size of the configuration, and to estimate its position on the surface of the earth. These four tasks are relatively independent of each other and range from the most topological and the least metric to the most metric. Only the last requires a knowledge of latitudes and longitudes. The emphasis in this essay is only on the configuration, without concern for positioning on the surface of the earth. The classical cartographic methods of estimating a configuration are based on triangulation (angle measurement), trilateration (distance measurement), or their combination in a traverse (distance and direction measurements). We can consider a configuration to be known when the locational components (coordinates) of all of its members are known. In an N dimensional space this requires the estimation of N components for each member of the configuration. N independent observations on each member are necessary to determine its N components. It is amusing to look at surveying books dated prior to circa 1750 to see how scrupulously and carefully they avoided using more than two observations relating to any one point. If one has such redundant observations they must be inconsistent, for every measurement contains error. The attitude has now completely changed, and it is considered good practice to have a minimum of $N + 1$ observations tying down each point, and even better to have much greater redundancies than this. The overmeasurement allows one to evaluate the internal consistency of the observational procedure, to detect blunders, and to evaluate the probable error at each point.

These ideas are nicely illustrated by the Brussels Distance List (Table 1), which dates from circa 1440. Several places are mentioned, but only eleven are related to each other by two or more observations. The separations of these eleven can be arrayed systematically (Table 2) and their locational components relative to each other can be computed by the standard survey adjustment procedure for a trilateration. The resulting configuration (Table 3 and Figure 1) includes the estimated standard error at each point.

TABLE 1
Brussels Distance List

De Wyenna ad Wratislaviam 48 miliaria. Item 24 ad Olomuntz. Item 32 ad Budam. Item 80 ad Venecias. Item 44 ad Saltzburgh. Item 36 ad Pataviam. Item 36 ad Pragam. Item 96 ad Basileam. Item 24 diete ad Mare Pontum ubi Danubius intrat mare.

De Saltzburgh ad Monacum 19 miliaria. Item 14 ad Pataviam. Item 24 ad Nurenbergam.

De Patavia ad Pragam 32 miliaria. Item 18 ad Ratisponam.

De Ingelstat ad Ulmam Item 9 ad Augustam. Item 11 ad Monacum. Item 7 ad Freising.

De Nurenberg 27 miliaria ad Erfordiam. Item 35 ad Liptz. Item 54 ad Maidburg. Item 74 ad Lubeck. Item 74 ad Wratislaviam. Item 100 ad Cracoviam. Item 36 ad Pragam. Item 68 ad Wyennam. Item 32 ad Pataviam. Item 14 ad Ratisponam. Item 34 ad Saltzburgh. Item 24 ad Monacum. Item 45 ad Basileam. Item 18 ad Ulmam. Item 24 ad Heidelberg. Item 32 ad Magunciam. Item 28 ad Francfordiam super Mogano. Item 60 ad Coloniam. Item 100 ad Parisius. Item 70 ad Jeng in Sabaudia. Item 20 ad Augustam. Item 13 ad Ingelstat.

SOURCE: Brussels, Bibliotheque Royale, MS. Inv. 1022-1047, f. 205v, col. 2; copied ca. A.D. 1440 by Paulus de Gherisheym.

TABLE 2
Distances from the Brussels Manuscript

	Nu	Vi	Pa	Sa	Pr	Mu	In	Br	Ba	Re	Au
Nurenberg	0	68	32	34	36	24	13	74	45	14	20
Vienna	68	0	36	44	36	—	—	48	96	—	—
Passau	32	36	0	14	32	—	—	—	—	18	—
Salzburg	34	44	14	0	—	19	—	—	—	—	—
Prague	36	36	32	—	0	—	—	—	—	—	—
Munich	24	—	—	19	—	0	11	—	—	—	—
Ingolstadt	13	—	—	—	—	11	0	—	—	—	9
Bratislava	74	48	—	—	—	—	—	0	—	—	—
Basel	45	96	—	—	—	—	—	—	0	—	—
Regensburg	14	—	18	—	—	—	—	—	—	0	—
Augsburg	20	—	—	—	—	—	9	—	—	—	0

Current optimal surveying design would attempt to distribute the observations so that the error ellipses are all circles of the same size (and small, of course), using a minimal, but adequately redundant, number of measurements.

The Brussels manuscript relates places in terms of distances, and it was compiled prior to the introduction of triangulation as the preferred (from circa 1500 to 1950) method of surveying. The choice of landmarks between which to observe distances or angles

TABLE 3
Trilateration Solution for Brussels Distance List

| | Solution Coordinates | | Residual Covariances | | | Percentage of Total MSE |
	x	y	δ_x	δ_y	δ_{xy}	(3.03737)
Nu	73.24	77.94	0.899	0.447	1.005	34.1
Vi	135.48	70.10	1.161	0.634	1.323	36.4
Pa	100.20	61.21	0.151	0.214	0.260	1.2
Sa	93.55	50.12	0.613	0.362	0.712	5.9
Pr	106.84	92.10	0.132	0.127	0.183	0.3
Mu	74.53	54.27	0.133	0.099	0.166	0.3
In	76.57	65.23	0.034	0.068	0.076	0.1
Br	130.25	25.42	0.965	0.133	0.164	16.8
Ba	39.58	50.61	0.789	0.378	0.875	4.9
Re	84.66	70.19	0.058	0.039	0.070	0.0
Au	70.43	58.45	0.041	0.114	0.121	0.1

is, in land surveying, usually fairly obvious, intervisibility of the stations being the dominant criterion. An additional rule of thumb is based on the so-called strength of figure. A glance at any surveying

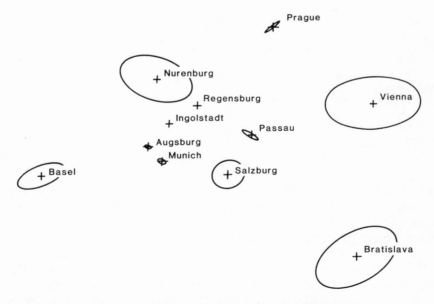

Figure 1. Solution configuration and standard errors.

text is recommended; it should convince one that a triangulation reduces the number of connected observations far below the possible r $(r - 1)/2$ links between r points. Local, that is, neighbor-to-neighbor, connections are emphasized. Even when the locations of the control points are known a priori, it is a nontrivial task to establish a net of triangles relating the points. This underscores the difficulty of establishing such a net when the task is to determine the location of the points. In land surveying, one further strengthens the sampling design by providing closure, as in a traverse that returns to its starting point or as in a net of triangles that is tied to previously determined (i.e., fixed) tie points. It is also usually estimated that the propagation error increases as the square root of the distance from these tie points. We now need to look at how these ideas mesh with those found in the estimation of proximity and preference.

The paper by Young, Sarle, and Hoffman presents an elegant computer-aided query system that allows for convergence toward an optimal sampling design as the answers begin to appear during an experiment. Thus it provides an on-line reconnoitering of the world. Perhaps more emphasis could be put on linking "neighboring" points as it becomes clearer during the sampling which points are near each other. Mild disappointments are the lack of standard error ellipses and the lack of a latitude/longitude graticule, state boundaries, and coastline, all on the U.S. map in Figure 7 in Chapter 1.2; and they do not give the data on which this figure is based. But interactive sampling designs seem to offer great economies, especially when they can be coupled to specific investigatory hypotheses. Can one design this so that the computer decides which hypotheses are worth exploration as the sampling continues?

Ian Spence has done a great deal to clarify some of the design possibilities in nonmetric multidimensional scaling. His notion (that the greater distances provide a framework to which the minor detail should be appended) is similar to modern land-surveying practice, wherein a widely spaced but highly precise geodetic survey forms the basis for local efforts. This result is perhaps slightly dependent on a quadratic error criterion that emphasizes the larger errors, assuming that the errors are proportional to the square roots of the estimated distances. Designs based on entries in the array seem somehow less ideal than the ones that one would establish if one knew the topology and geometric arrangements of the set members in advance, particularly since the ordering of the elements in the array is completely arbitrary. The problem, of course, is that one cannot set out an optimal triangulation until one knows which pairs are neighbors. Some interesting criteria for redundancy are cited by Spence, and these obviously depend on the dimensionality of the space in which one expects a solution. It might also be desirable to check violations of the triangular inequality or the symmetry of the proximity estimates, since these are topologically fundamental criteria for the hypothesis of a geometric model. Some greater emphasis in the design choice might be put on evenly distributing the error over the entire space, rather than using only a global error measure.

Spector and Rivizzigno seem to have done something very curious. Here is a case in which a close approximation to the final configuration immediately suggests itself. They do not take advantage of this and have used the array to prepare the design. They do not even show what this design looks like on the geographical map; however, one could construct this from the data they give. These seem to be a pair of geographers who are ignorant of the geographical tradition. There are no standard error estimates on their maps, nor do their design criteria seem cognizant of any such possibilities. What they offer is a largish number of experiments, which may be suggestive when one lacks an a priori estimate of the form of the configuration. But should one do any experiment in this kind of a vacuum? An investigator should have some ideas. To pretend that he or she does not is "objectivity" taken to a ridiculous extreme. They rightly point out, although it seems rather obvious, that similar numerical values of the "stress" (the nonmetric equivalent of the surveyor's root-means-square-error) can be obtained for radically different configurations from the same set of data. This is almost never a problem in land surveying, since one always has a good starting configuration for the trilateration or triangulation iterations.

The paper by Paul Isaac discusses many of the same issues that I have covered from a slightly different point of view. He does some numerical experiments to evaluate a particular way of choosing pairs to be compared. The specific configurations that he attempts to recover seem to me to be capriciously arbitrary. But I know of no theory that would lead to a better selection; surely this must be subject matter dependent.

Arabie and Soli address a rather different topic. Basically, I find their point of view agreeable, for I have always felt that one of the more interesting results of a multidimensional scaling analysis is contained in the Shepard diagram. Curiously, many of the recent computer programs omit this graphical output. Thus, if one puts in data from an ellipsoid surface (such as the earth), a hint of this should show in the graph relating input dissimilarities to output distances. One could (should?) then redo the analysis with a Riemann metric instead of an Euclidean one. I cannot comment meaningfully on the linguistic data that they analyze, but I suspect that similar preprocessing of geographical data would be useful.

REFERENCES

1. Beazley, C. R., 1895. *Prince Henry the Navigator*. London: G. P. Putnam's Sons.
2. Bjerhammer, A., 1972. *Theory of Errors and Generalized Inverses*. Amsterdam: Elsevier.
3. Bossler, J. D., and E. Grafarend, 1973. "Optimal Design of Geodetic Nets," *Journal of Geophysical Research*, 78, 5887.
4. Brown, L. A., 1949. *The Story of Maps*. Boston: Little, Brown and Company.
5. Dilke, O., 1971. *The Roman Land Surveyors*. Newton Abbot: David and Charles.
6. Golledge, R. G., and G. Rushton, 1972. *Multidimensional Scaling: Review and Geographical Applications*. Washington, D.C.: Association of American Geographers, Commission on College Geography, Technical Paper #100.
7. Grafarend, E., 1974. "Optimization of Geodetic Networks," *Bolletino di Geodesia e Scienze Affini*, 33, 351.
8. Hirvonen, R. A., 1971. *Adjustment by Least Squares in Geodesy and Photogrammetry*. New York: Frederick Unger.
9. Kendall, D. G., 1971. "Construction of Maps from 'Odd' Bits of Information," *Nature*, 231, 158.
10. Lawson, C., 1972. "Transforming Triangulations," *Discrete Mathematics*, 3 (4), 365.
11. Mikhail, E. M., 1976. *Observations and Least Squares*. New York: Thomas Y. Crowell Company.
12. Pieszko, H., 1970. "Multidimensional Scaling in Riemann Space." Ph.D. dissertation, Department of Psychology, University of Illinois, Urbana, Ill.
13. Price, A. G. (ed.), 1971. *The Explorations of Captain James Cook in the Pacific*. New York: Limited Editions Club.
14. Richeson, A. W., 1966. *English Land Measuring to 1800*. Cambridge, Mass.: Society for the History of Technology.
15. Shepard, R. N., 1962, "Analysis of Proximities," *Psychometrika*, 27, 125 and 219.
16. Taylor, E. G. R., 1971. *The Haven Finding Art*. London: Hollis and Carter.
17. Tobler, W. R., 1976. "The Geometry of Mental Maps." In R. G. Golledge and G. Rushton (eds.), *Spatial Choice and Spatial Behavior*, pp. 69-81. Columbus, Ohio: Ohio State University Press.
18. Tobler, W. R., 1977. "Numerical Approaches to Map Projections." In E. Kretschmer (ed.), *Studies in Theoretical Cartography*, pp. 51-64. Vienna: Deuticke Verlag.
19. Wolf, P., 1969. "Horizontal Position Adjustment," *Surveying and Mapping*, 29, 635.

Chapter 1.2 Interactively Ordering the Similarities among a Large Set of Stimuli

Forrest W. Young, Cynthia H. Null,
Warren S. Sarle, and *Donna L. Hoffman*

When using a multidimensional scaling procedure, most researchers collect data by having subjects rate the similarity of all possible pairs of stimuli. When the number of stimuli is large (say 25), the number of pairs is very large (300), and the task is extremely arduous. Since using a large number of stimuli is often desirable and is necessary if one expects a solution with more than three dimensions [10], some way of limiting the size of the task is of great interest.

Many researchers have avoided the large number of pairs required by a large number of stimuli by using a sorting or grouping task instead of interstimulus similarity judgments [e.g., 2 and 6]. A similarity matrix is derived from the sorting information. One may question whether the derived matrix adequately approximates the subject's perception of the interstimulus distances. This method may be difficult to use in some situations. For example, it would be very difficult to sort auditory stimuli, since they appear in time and cannot be physically sorted into groups.

If the researcher collects data using ratings of pairs and wishes to investigate a large number of stimuli, then the number of pairs must be reduced. Spence [7, 8, 9] has proposed using a cyclic design to choose a subset of the stimulus pairs. He has demonstrated

that gathering data for approximately 60% of the pairs yields results comparable to other procedures. He has found that a solution
space is best defined by its maximally distant pairs. Use of a cyclic
design does not guarantee the presentation of these pairs unless
the experimenter has prior knowledge of the stimuli and chooses
these pairs to appear in the cyclic subsets.

Young and Cliff [11] have developed an interactive procedure
(ISIS) that chooses the subset of pairs to be presented on the basis
of the subject's previous responses. The ISIS procedure determines
the maximally distant stimuli, so that prior knowledge is not necessary to obtain a well-defined space. The procedure requires the
subject to make judgments on 25% to 45% of the pairs. This technique has been favorably evaluated with both empirical and Monte
Carlo procedures [1, 3, 4]. However, the procedure assumes that
the matrix (O) of observations is Euclidean.

The interactive similarity ordering (ISO) system allows the experimenter to obtain ordinal similarity data for at least as large a
set of stimuli as do the procedures just mentioned. This system
can be used to collect data that are symmetric, asymmetric, or rectangular. ISO does not use the Euclidean model to help gather
data; rather, it uses transitivity, a much weaker assumption underlying all non-Euclidean (and Euclidean) metrics. The ISO system
has two important advantages: (1) the subject makes a very simple
judgment, and (2) the number of judgments is interactively minimized.

ISO EMPIRICAL PARADIGMS

Definitions

The data collected using the ISO system can be described as
two-way data, where the number of ways corresponds to the number of experimental conditions. Two-way data can be either square
or rectangular. Square data involve the Cartesian product ($i \times i$) of
a set of stimuli i with itself. For example, a researcher could ask a
subject to rate the similarity of twenty-five comedians to each
other. These data would be square two-way data and also symmetric. The distinguishing feature of a symmetric square matrix O of
observations is that the entries $o_{ij} = o_{ji}$ for all pairs of elements.
(See Figure 1.) Square data can also be asymmetric. Usually, asymmetric data are also row conditional; that is, the similarities in the
ith row are relative to the ith stimulus. For example, an experimenter could ask the subject to rank order twenty-four of the

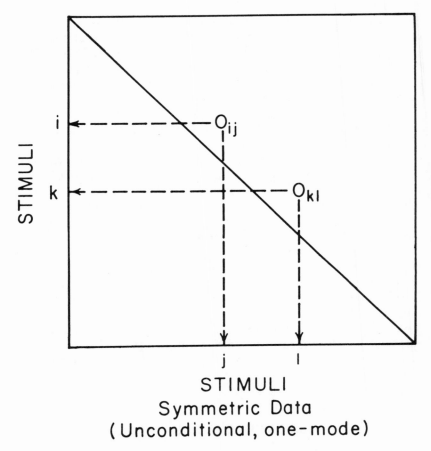

STIMULI
Symmetric Data
(Unconditional, one-mode)

Figure 1. Schematic of symmetric data matrix.

comedians in relation to the twenty-fifth. The subject would then repeat this ranking task twenty-four more times, with each of the comedians as the standard. These data are square because the matrix is comedian by comedian but are asymmetric because $o_{ij} \neq o_{ji}$. (See Figure 2.) Finally, two-way data can be rectangular. Such data involve the Cartesian product ($i \times j$) of two sets, i and j. (See Figure 3.) An example of rectangular data would be when a researcher has a subject rate twenty musical timbres on each of eight adjective scales. The experimenter might have the subject rate how "brilliant" each timbre was, how "hollow" each timbre was, and so on.

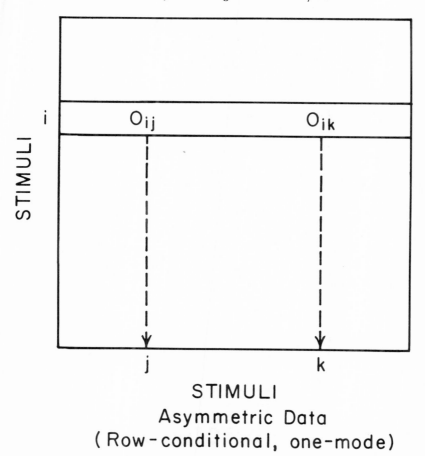

STIMULI
Asymmetric Data
(Row - conditional, one-mode)

Figure 2. Schematic of asymmetric data matrix.

The Task

When collecting square, symmetric data about n stimuli, ISO presents the subject with the following judgment situation: A list of from 2 to m pairs of stimuli is presented to the subject, who is asked to select the pair whose elements he or she judges to be most similar. The subject is repeatedly presented with such lists until a rank order of $p < n(n-1)/2$ pairs is obtained. Figure 4 presents examples of two trials in which the stimuli are American cities. The maximum length of the list presented, m, can vary from 2 through $n(n-1)/2$. From one presentation to the next, only 1

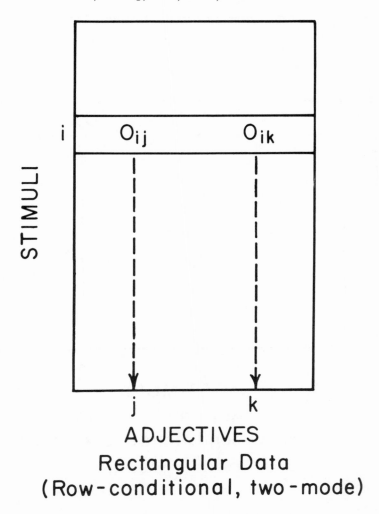

Figure 3. Schematic of rectangular data matrix.

pair changes, so that the task is quite simple. This method of collection is only appropriate when the number of stimuli is relatively small (say, under 12), unless an order for all pairs is not desired. Thus, we will not consider this method in the remainder of this paper.

In the conditional rank order situation (square, asymmetric data), one of the n stimuli is designated the "standard" stimulus

WHICH OF THESE PAIRS IS CLOSEST?

   2 MEMPHIS
     SALT LAKE CITY

   3 LOS ANGELES
     SALT LAKE CITY

   4 BOSTON
     WASHINGTON

   YOUR RESPONSE: 4

WHICH OF THESE PAIRS IS CLOSEST?

   2 MEMPHIS
     SALT LAKE CITY

   3 LOS ANGELES
     SALT LAKE CITY

   YOUR RESPONSE: 3

Figure 4. Example of subject judgment task using symmetric experimental paradigm (maximum list length of three).

and the remaining $n - 1$, the "comparison" stimuli. A list of from 2 to m comparison stimuli is presented to the subject, who is asked to select the one he or she judges to be most like the standard. The subject is repeatedly presented with such lists until the rank order of the similarity of p of the $n - 1$ comparison stimuli to the standard is determined ($p \leqslant n - 1$). Usually, only 1 stimulus in the list changes from one presentation to the next, thus simplifying the task. The maximum length of the list of comparison stimuli is set by the experimenter and can vary from 2 through $n - 1$. In addition, the maximum number of stimuli to be ordered, p, is also set by the experimenter. In Figure 5 are three trials in which the maximum list length is 4 and the stimuli are again cities. For this set of trials Boston was chosen as the standard.

Note that the list length m is the maximum number to be presented on any trial. Therefore, the subject may see as few as 2

WHICH OF THE FOLLOWING CITIES IS CLOSEST TO: BOSTON

    3 WASHINGTON

    4 LOS ANGELES

  YOUR RESPONSE: 3

WHICH OF THE FOLLOWING CITIES IS CLOSEST TO: BOSTON

    1 MEMPHIS

    2 NEW ORLEANS

    3 CHAPEL HILL

    4 WASHINGTON

  YOUR RESPONSE: 4

WHICH OF THE FOLLOWING CITIES IS CLOSEST TO: BOSTON

    1 MEMPHIS

    2 NEW ORLEANS

    3 CHAPEL HILL

    4 LOS ANGELES

  YOUR RESPONSE: 3

Figure 5. Example of subject judgment task using asymmetric experimental paradigm (maximum list length of four).

comparison stimuli on some trials, as happens in the example. The maximum length of the comparison lists will affect the total number of judgments a subject must make for each standard and the difficulty of each judgment.

After completing one conditional rank order, ISO selects a second standard and has the subject make selections from lists of comparison stimuli, thus constructing a second conditional rank order. The order in which the standards are selected can be chosen by the experimenter or can be randomly determined by ISO. This procedure is repeated until $c \leqslant n$ conditional rank orders have been completed, where c is also set by the experimenter and can vary from 1 to n. This results in an asymmetric matrix of two-way similarity data.

This procedure is appropriate for a relatively large $(n > 30)$ number of stimuli, for a variety of reasons. First, the subject is rank ordering $n - 1$ items and not $n(n - 1)/2$ pairs. Second, not all standards must be rated by the subjects. Third, partial orders of the comparison stimuli can be obtained. For a partial order the comparison stimuli are divided into sublists and the subjects must determine a rank order for each of these sublists but not for the complete list. This reduces the number of judgments a subject must make for each standard. All of these judgment-saving options are available in the ISO system and can be used to shorten the task at the experimenter's discretion.

We should also point out that the maximum length of the list of comparison stimuli affects the total number of judgments, as well as the difficulty of the judgment. Specifically, fewer judgments are required to order a set of p comparison stimuli with long lists. For example, p comparison stimuli can be ordered with $p - 1$ judgments when the maximum list length is p. On the other hand, the number of judgments is approximately $p(\log_2(p - 1))$ when the list length is only 2. Thus, if $p = 16$ is the number of comparison stimuli and $m = 16$ is the maximum list length, the subject makes 15 judgments. However, with $p = 16$ and $m = 2$, the subject makes 48 judgments. An intermediate number of judgments is needed for intermediate list lengths. Of course, the longer the list, the more difficult and time-consuming each judgment becomes. It takes a long time to decide which of 16 comparison stimuli is most like the standard but a very short time to decide which of 2 comparison stimuli is more like the standard. Thus, there may be some intermediate list length that requires the least time to order the p comparison stimuli.

To obtain rating scale (rectangular) data, the researcher presents the subject with an adjective as the "standard" and p of the n stimuli to be ordered with respect to the standard. As in the conditional rank order task, a list of from 2 to m comparison stimuli $(m \leqslant p \leqslant n)$ is presented to the subject, who is asked to choose the stimulus that is best described by the adjective or is most like the adjective, and so on. In the example (Figure 6), the subject is asked to indicate which of the cities is most familiar to him or her. Lists of comparison stimuli are presented, usually with one item changing per presentation, until a rank order for the stimuli with regard

WHICH OF THE FOLLOWING CITIES IS MOST FAMILIAR TO YOU:

    2 WASHINGTON

    4 BOSTON

  YOUR RESPONSE: 2

WHICH OF THE FOLLOWING CITIES IS MOST FAMILIAR TO YOU:

    1 CHAPEL HILL

    2 SALT LAKE CITY

    3 DALLAS

    4 LOS ANGELES

  YOUR RESPONSE: 1

WHICH OF THE FOLLOWING CITIES IS MOST FAMILIAR TO YOU:

    1 DENVER

    2 SALT LAKE CITY

    3 DALLAS

    4 LOS ANGELES

  YOUR RESPONSE: 4

Figure 6. Example of subject judgment task using rectangular experimental paradigm (maximum list length of four).

to the standard is obtained. This procedure is repeated for each adjective. All of the labor-saving options in the conditional rank order task are available.

Algorithm

Computers are often called upon to arrange a set of items, usually numbers, into order. Computer scientists call this process "sorting." Sorting has been extensively studied to find the best possible algorithms. Sorting procedures frequently order items by making comparisons; for example, which of two numbers is larger, or which or two words comes first in alphabetical order. Any sorting algorithm that uses comparisons can be adapted to interactively order stimuli by simply asking a person to make the comparison rather than having the computer do it. In fact, humans have a definite advantage over the computer in that they can compare several stimuli at once to pick out the largest, closest, most familiar, and so on, while computers can compare only two things at a time.

In choosing an algorithm for interactive stimulus ordering, three criteria should be considered: (1) the number of comparisons the subject makes should be as small as possible; (2) the comparisons should be as easy as possible; and (3) it should be possible to take advantage of the subject's ability to make multiple comparisons. An algorithm that does very well in satisfying these three criteria is the m-way merge sort. The letter m refers to the maximum number of stimuli involved in a comparison, for example, the list length in Figures 4 and 5. An m-way merge sort applied to n stimuli requires at most $n(\log_m(n - 1))$ comparisons, where the logarithm is taken to the base m and should be rounded upward when it does not come out even. It is mathematically impossible to improve substantially on this number of comparisons. The comparisons required are like those in Figures 4 and 5 and are very easy for the subject to make. For the vast majority of comparisons, the list differs from the previous list in only one stimulus. This property of the m-way merge sort makes the transition from one list to the next very easy. The comparison list can, of course, have as many items as the experimenter desires.

The m-way merge sort is based on the idea of taking m lists of stimuli, each of which has been separately ordered, and combining these lists into one completely ordered list. When the algorithm begins, each stimulus is considered a separate list by itself. Then, the computer repeatedly merges these lists into larger and larger lists until all the stimuli are in a single ordered list.

A two-way merge sort with sixteen cities ordered according to familiarity to a hypothetical subject is illustrated in Table 1. Column 1 shows the cities in random order. The horizontal lines separate lists that have been ordered. At the first stage (i.e., column 1), each city constitutes a separate list. The computer first presents the subject with the cities Chapel Hill and Denver, using a display such as in Figure 6. The subject decides that Chapel Hill is more familiar. Then, Chapel Hill and Denver can be combined into an ordered list, shown at the top of column 2. The computer next presents the cities San Francisco and Chicago, and the subject

TABLE 1
Two-way Merge Sort Based on Familiarity

Column 1	Column 2	Column 3	Column 4	Column 5
Denver	Chapel Hill	Chapel Hill	Chapel Hill	Chapel Hill
Chapel Hill	Denver	Chicago	New York	Washington, D.C.
San Francisco	Chicago	Denver	Boston	Atlanta
Chicago	San Francisco	San Francisco	Chicago	Miami
Memphis	Boston	New York	Denver	New York
Boston	Memphis	Boston	Memphis	Boston
Dallas	New York	Memphis	Dallas	Chicago
New York	Dallas	Dallas	San Francisco	Philadelphia
Altanta	Atlanta	Washington, D.C.	Washington, D.C.	New Orleans
Philadelphia	Philadelphia	Atlanta	Atlanta	Denver
Washington, D.C.	Washington, D.C.	Philadelphia	Miami	Memphis
Salt Lake City	Salt Lake City	Salt Lake City	Philadelphia	Dallas
Miami	Miami	Miami	New Orleans	Salt Lake City
New Orleans	New Orleans	New Orleans	Salt Lake City	Los Angeles
Los Angeles	Los Angeles	Los Angeles	Los Angeles	San Francisco
Seattle	Seattle	Seattle	Seattle	Seattle

chooses Chicago as more familiar. This decision results in the second ordered list in column 2. In a similar way, the rest of the cities in column 1 are merged into ordered lists containing two cities each, as shown in column 2.

Next, the computer must merge the first two lists, with two cities each, in column 2. The subject is shown the first city in each of these lists, Chapel Hill and Chicago, and he or she decides that Chapel Hill is more familiar. The computer removes Chapel Hill from the comparison list and replaces it with the next city from that list, Denver. The subject says that Chicago is more familiar than Denver, so Chicago is removed from the comparison list and replaced by San Francisco. The subject chooses Denver as more familiar than San Francisco. At this point, the computer can deduce the complete order for these four cities, which is shown as the first ordered list in column 3. The computer can proceed to merge the list containing Boston and Memphis with that having New York and Dallas, yielding the second ordered list in column 3. The last two lists in column 3 are produced in a similar fashion. The merging process is continued to generate column 4 from column 3 and the complete order in column 5 from column 4.

Actually, the merges do not occur in the exact order given above, but the description is easier to understand than what the computer actually does. Furthermore, the description yields results that are the same as the computer's. There are complications the computer must handle when the number of stimuli to be ordered is not an exact power of the list length. The merges are arranged to minimize the number of comparisons required.

Program

The ISO system has been programmed in FORTRAN-4 PLUS for a PDP-11/45 with an RSX-11D operating system and CRT terminal. This program will collect data for any of the paradigms mentioned above. It will also collect data when the task is to insert into a list of comparison stimuli (using a generalized multipivot quicksort algorithm) as opposed to picking from a comparison list as described here. The program, in addition to collecting similarity data, determines the number of judgments per order and the time per judgment. For the asymmetric and rectangular paradigms, the subject may interrupt a data-collection session between standards and return at a later time to complete the task. This allows the subject to stop when tired and return when fresh, thus improving judgment reliability and validity. It also enables collection of data over several sessions, thus further increasing the potential number of stimuli.

EXPERIMENT 1

For investigating the effectiveness of the ISO system, data concerning sixteen stimuli were collected. The asymmetric paradigm was used. The effect of comparison list length was assessed on two dependent variables: the number of judgments the subject made per standard and the length of time for each judgment. Comparison lists (m) were either two, eight, or fifteen stimuli long. The use of partial orders was also investigated. Instead of ranking the complete list for each standard, half of the subjects in the list length conditions of two and eight rank ordered two sublists. Thus, there were five experimental conditions, complete orders of the stimuli for presentation list lengths of two, eight, and fifteen, and partial orders (or separately ranking each half of the stimuli) for list lengths of two and eight.

Subjects

Forty-five undergraduate introductory psychology students at the University of North Carolina at Chapel Hill served as subjects. They received course credit for their participation. The subjects were randomly assigned to one of the five experimental conditions. The subjects were from North Carolina.

Stimuli

Sixteen American cities were used as stimuli: Atlanta, Boston, Chapel Hill, Chicago, Dallas, Denver, Los Angeles, Memphis, Miami, New Orleans, New York, Philadelphia, Salt Lake City, San Francisco, Seattle, and Washington, D.C.

Apparatus

The stimuli were presented to the subject via the ISO system described above. The display terminal operated at 9600 BAUD.

Procedure

The subject sat in front of the computer terminal; and, after giving the subject brief instructions on using the keyboard (and after a warm-up task), the experiment began. The subject was instructed, via the terminal, to rank order the comparison cities in terms of their perceived distances from the standard city. The subject produced one or two rank orders per standard stimulus. Each order contained either seven, eight, or fifteen cities, depending on the experimental condition. The subject was presented with eight standards.

Results

A summary of the results is contained in Table 2. As expected, the average time per judgment increased, and the number of judgments decreased as the length of the comparison list was increased. When we multiply these two numbers together and then multiply the result by two for the partial-order conditions, we obtain the average time per standard. From the table we can see that the complete-order conditions required much more time than the partial-order conditions, with the partial-eight condition requiring 2.2 minutes per standard and the complete conditions requiring 2.8 to 3.2 minutes per standard. From these results it is estimated that subjects can judge eighteen to twenty comparison stimuli during an hour with the complete order tasks and as many as twenty-seven or thirty in the partial condition with list length of eight. Finally, from these results it can be estimated that with thirty-three stimuli a subject in a partial-order task involving four sublists per standard (with eight stimuli per sublist) could complete sixteen standards in 75 minutes.

The multidimensional scaling solution for the complete-order condition appears in Figure 7. Solutions for each list length are not separately presented, since list length did not affect solution quality. The solution in Figure 7 can be thought of as equivalent

TABLE 2
Evaluation of the ISO System for the Asymmetric Experimental Paradigm

List Length Type of Ordering	2 Complete	2 Partial	8 Complete	8 Partial	15 Complete
Average time per judgment (in seconds)	4.707	5.519	9.169	9.983	12.348
Standard deviation (in seconds)	2.076	1.813	2.805	6.091	4.572
Average number of judgments per order	41.284	14.589	21.21	6.5	14.0
Standard deviation	2.972	2.022	0.0721	0.0	0.0
Average number of minutes per standard	3.239	2.684	3.241	2.166	2.881
Number of standards per hour	18	22	18	27	21
Number of orders	67	180	62	129	56
Number of subjects	9	10	8	10	8

Figure 7. Multidimensional scaling solution for asymmetric data from the complete-order condition (sixteen stimuli).

to a map of the United States that has been drawn on a rubbery surface and then stretched and shrunk in various ways (except for the location of San Francisco and the reversing in the east-west direction of Chapel Hill and Atlanta). Previous results indicate that only this type of equivalence to a U.S. map should be expected from subjective distance judgments. It can be argued that the subjects can more accurately judge distances in the part of the country where they reside and that the distant locations will lump together. Thus, the distances between western cities should be decreased relative to true distances for residents of North Carolina.

The MDS solution for the partial-order asymmetric data was very different from a map of the United States. This was due to a mistake in the design of the experiment [12]. This simple design error invalidated the results of the MDS solution but had no detrimental effect on the information presented in Table 2. Due to this error, a second experiment was designed for the partial-order asymmetric paradigm.

EXPERIMENT 2

Subjects

Twenty-three undergraduate introductory psychology students at the University of North Carolina at Chapel Hill served as subjects. They received course credit for their participation. The subjects were from North Carolina.

Stimuli

Twenty-nine American cities were used as stimuli: Albuquerque, Atlanta, Billings, Boston, Chapel Hill, Chicago, Cincinnati, Dallas, Denver, Kansas City, Los Angeles, Memphis, Miami, Minneapolis, Montgomery, Nashville, New Orleans, New York, Oklahoma City, Philadelphia, Phoenix, Pittsburgh, Portland, Rochester, Salt Lake City, San Diego, San Francisco, Seattle, and Washington, D.C.

Apparatus

Stimuli were presented via the same apparatus as in the first experiment.

Procedure

One of the twenty-nine stimuli was designated the standard stimulus and the remaining twenty-eight, the comparison stimuli. The twenty-eight comparison stimuli were then randomly subdivided into four lists of seven stimuli each. The four lists were mutually exclusive and exhaustive.

The subject was presented with the standard and one of the lists of seven comparison stimuli. The subject was then asked to choose the one stimulus from the list of seven that he or she judged to be the closest to the standard. This judgment was repeated until all seven stimuli in the comparison list were ordered. Then, a second list of seven comparison stimuli was selected at random, and the procedure was repeated with the same standard stimulus. This procedure was then repeated for the third and fourth lists of seven comparison stimuli.

Each subject was presented with six to eight different randomly selected standard stimuli. Four comparison lists of seven stimuli were formed for each standard. The composition of the lists was determined anew for each standard (by random selection), thereby ensuring that the lists were different for each standard. This procedure ensured that all stimuli were "connected"; that is, there were data to determine the distances between all pairs of stimuli. (In the previous partial-order experiment, the stimuli were not completely connected, thereby invalidating the MDS structure.) The experimental session required 60 minutes of the subject's time.

Results and Discussion

The two-dimensional space appears in Figure 8. As we can see, the structure of the MDS solution is highly similar to a map of the United States. This solution has the same general characteristics as the solution from the complete-order condition in the previous experiment, even though (1) there were two times the number of stimuli, (2) there was the same number of subjects, and (3) the subjects spent the same amount of time at the task as in the previous experiment.

The number by each stimulus is a squared correlation index, and it is interpreted as the proportion of variance of the transformed judgments (disparities) accounted for by the model displayed in Figure 8 (the interpoint distances). Higher values reflect a better fit of the model to the data. They also indicate the amount of agreement between subjects for each stimulus. Note that a squared correlation of 1.00 means that all subjects responded in the same way for that stimulus.

Inspection of the squared correlations for each stimulus reveals an interesting pattern. The dashed line divides the solution into two sections. This line, which roughly follows the Mississippi River, separates higher correlations from lower correlations. Most of the values to the left of the Mississippi are less than 0.985, while most to the right are greater than 0.985. This division reflects the fact that the subjects came from North Carolina and lends further credence to the argument that subjects can more accurately judge distances between cities they are familiar with.

CONCLUSIONS

We conclude that ISO can gather data concerning a moderately large number of stimuli without wearing out the subject and that

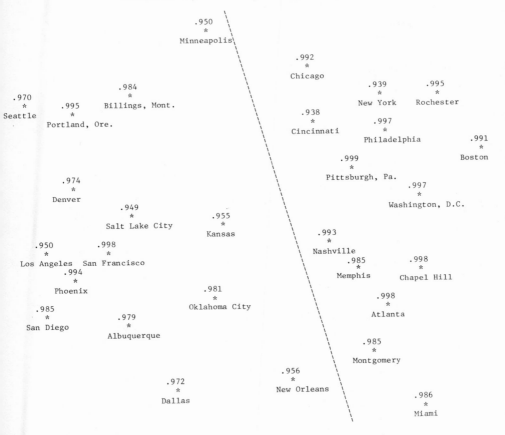

Figure 8. Multidimensional scaling solution for asymmetric data from the partial-order condition (twenty-nine stimuli).

data are of high enough quality to permit meaningful multidimensional scaling. However, we look on the investigations reported here as providing preliminary results that only enable us to reach tentative conclusions. Further investigations are being performed to determine, by Monte Carlo techniques, the effect of (1) accidentally selecting the "wrong" stimulus from the list of comparison stimuli and (2) the effect of misperceiving the similarity of comparison stimuli to the standard. These investigations are determining the effect of such errors on the derived configuration and on the validity of the transitivity assumption. Finally, we anticipate using ISO to gather both similarity and attribute data about stimuli for which there is no physical structure but for which previous investigations have revealed well-defined cognitive structure. Such an investigation will provide a more difficult evaluation hurdle than provided by the current study.

REFERENCES

1. Baker, R. F., and F. W. Young, 1974. *Interactive Scaling with Individual Subjects II: An Improved System.* Chapel Hill, N.C.: University of North Carolina, L. L. Thurstone Psychometric Laboratory Report #132.
2. Boorman, S. A., and P. Arabie, 1972. "Structural Measures and the Method of Sorting." In R. N. Shepard, A. K. Romney, and S. B. Nerlove (eds.), *Multidimensional Scaling: Theory and Applications in the Behavioral Sciences* (vol. 1), pp. 225-49. New York: Seminar Press.
3. Girard, R. A., and N. Cliff, 1976. "A Monte Carlo Evaluation of Interactive Multidimensional Scaling," *Psychometrika*, 41, 43-64.
4. Hamer, R., 1978. "Nonmetric Interactive Scaling with Multiple Subjects." Ph.D. dissertation, Department of Psychology, University of North Carolina, Chapel Hill.
5. Knuth, D. E., 1973. "Sorting and Searching." *The Art of Computer Programming* (vol. 3). Reading, Mass.: Addison-Wesley.
6. Rosenberg, S., and A. Sedlak, 1972. "Structural Representations of Perceived Personality Trait Relationships." In A. K. Romney, R. N. Shepard, and S. B. Nerlove (eds.), *Multidimensional Scaling: Theory and Applications in the Behavioral Sciences* (vol. 2), pp. 133-62. New York: Seminar Press.
7. Spence, I., 1981. "Incomplete Experimental Designs for Multidimensional Scaling." In R. G. Golledge and J. N. Rayner (eds.), *Proximity and Preference: Problems in the Multidimensional Analysis of Large Data Sets.* Minneapolis: University of Minnesota Press.
8. Spence, I., and D. W. Domoney, 1974. "Single Subject Incomplete Designs for Nonmetric Multidimensional Scaling," *Psychometrika*, 39, 469-90.
9. Spence, I., and H. Janssen, 1974. "Cyclic Incomplete Designs as an Alternative to ISIS Models in Nonmetric Scaling," a paper presented at Psychometric Society Meetings, Stanford University, 1974.
10. Young, F. W., 1970. "Nonmetric Multidimensional Scaling: Recovery of Metric Information," *Psychometrika*, 35, 455-73.
11. Young, F. W., and N. Cliff, 1972. "Interactive Scaling with Individual Subjects," *Psychometrika*, 37, 385-415.
12. Young, F. W., C. H. Null, and W. Sarle, 1978. "Interactive Similarity Ordering." *Behavior Research Methods and Instrumentation*, 10, 273-80.

Chapter 1.3 Incomplete Experimental Designs for Multidimensional Scaling

Ian Spence

The major problem that faces a researcher who desires to use multidimensional scaling procedures with large numbers of stimulus objects is the difficulty, or undesirability, of presenting a subject with all possible pairs of the stimuli. This number rises almost as the square of the number of objects; for example, with 10 or 20 stimuli the subject has to judge 45 or 190 pairs, respectively, whereas with 40 or 50 stimuli the number of judgments required rises to 780 or 1225. Clearly, there comes a point when the patience of even the most highly motivated subject is exhausted, not to mention the questionable quality of the resulting data. With moderately large values of n (the number of stimuli) the pairwise judgment task becomes onerous, and certainly, when n exceeds 60 or 70, using the complete paired comparison task is almost always out of the question.

A number of possible alternatives have recently been proposed for dealing with large stimulus sets. Most published applications involving large n's have sidestepped the problem by using techniques that do not require paired comparison judgments; these applications usually involve sorting or grouping tasks of various

The research on which this chapter is based was supported by grant A8351 from the National Research Council of Canada. I am grateful to Henry Janssen for collecting and analyzing the data reported in the empirical example.

kinds [e.g., 2, 6, 15, 16]. Subsequent to the sorting task, a proximity matrix may be derived, and then multidimensional scaling is performed with the complete matrix. Whether this matrix adequately represents the subject's perception of the pairwise interstimulus relations is a moot point. Indeed, although, unfortunately, no serious reservations have appeared in print, there are some researchers, highly experienced in the use of such techniques, who advocate great caution and careful examination of the results obtained by using these methods. Furthermore, there are some situations in which these procedures simply cannot be used; imagine a geographer interested in constructing mental maps asking a subject to sort places into homogeneous categories! Such methods are not well suited to the examination of individual perceptual spaces.

When an experimenter wishes the subject to make pairwise judgments, the only solution is to present some manageable subset of the possible $n(n - 1)/2$ pairs. Young and Cliff [23] have devised a computer program that obtains a subset of the judgments from a subject in real time. This subset is not determined in advance but is selected by the interactive scaling program on the basis of the subject's previous responses. The procedure appears to work very well [1, 7], but it does suffer from one salient disadvantage: the experimenter requires access to time-sharing facilities and terminals. Moreover, group testing becomes difficult, if not impossible, and there are constraints on the types of stimuli that can be presented via the terminal, as well as constraints on the subject's mode of response; and, obviously, experimentation in the field becomes a problem of nontrivial proportions.

Spence and Domoney [19] have proposed that a subset of the stimulus pairs be chosen in advance according to an explicit experimental design. This has many of the advantages of the interactive approach but eliminates the difficulties already mentioned. The real problem remaining is how to choose a good subset of the stimuli, both in terms of its design and its size. Random designs, overlapping clique designs, and two different cyclic designs were examined by Spence and Domoney [19]. Random selection of the stimulus pairs was found, perhaps surprisingly, to be quite effective, with performance not much inferior to high-efficiency cyclic designs. The overlapping clique designs [20, p. 192, Figure 5 (c)], did not perform well, and Spence and Domoney [19] conjectured that this would also be true for any design that was strongly locally connected. They also found that of the two cyclic designs used, the one with the higher efficiency outperformed all the other designs; this led to the conjecture that connectedness — as measured by efficiency, mean resistance, or number of triangles — was an important and desirable property of an incomplete design. Before describing an experiment that examines this contention, it is perhaps appropriate to review some of the construction details and properties of cyclic designs.

CYCLIC DESIGNS

Methods of constructing cyclic paired comparison designs have been described by David [5], John, Wolock, and David [10], and Clatworthy [4], among others. However, since most applied researchers are probably not familiar with these designs, a simple method, which has been proposed by David [5], will be presented, and the procedure will be illustrated by a small example.

1. Number the stimuli 0, 1, 2, . . . , $n - 1$.
2. Consider the *cyclic sets*
   $\{s\}$: $(0,s)$ $(1,s+1)$ . . . $(t,s+1)$ . . . $(n-1,s+n-1)$ with reduction modulo n where necessary (i.e., the remainder after division by n).
   For example, with $n = 10$
   $\{3\}$: $(0,3)$ $(1,4)$ . . . $(4,7)$ . . . $(9,2)$
   Each cyclic set contains n pairs of stimuli. Each stimulus appears equally often. The pairs are usually unique, but in certain cases may not be. The set is connected if s is relatively prime to n. [For further details, see 5.] Figure 1 shows the graph of cyclic set $\{3\}$.
3. Combine the desired number of cyclic sets to form a *cyclic design*.
   For example, with $n = 10$ say
   Design $\{1, 3\} = \{1\} + \{3\}$

In Figure 1, a small example for the case of 10 stimuli is presented; two designs with 20 of the possible 45 comparisons are shown. This example illustrates the fact that, even though different cyclic designs may have the same number of pairs, their efficiencies can differ. It should be noted also that cyclic set $\{2\}$ is not connected — 2 is not relatively prime to 10.

The efficiency of a cyclic design, in analysis of variance terms, can be defined as the ratio of the average between treatment variance for the incomplete design to the average between treatment variance for the complete design [12]; this turns out to be:

$$E = \frac{(n-1)^2}{nrb}$$

$$\text{with } b = \sum_{\ell=1}^{m} \left(\frac{1}{\lambda_\ell}\right) + \frac{1}{2\lambda_{\frac12 n}} \cdot (n-1) \bmod 2$$

$$\text{and } \lambda_\ell = \frac{r}{2} - \sum_{k=1}^{m} g_k \cos 2\pi(k\ell/n) - 1/2(-1)^\ell g_{\frac12 n} \cdot (n-1) \bmod 2$$

where $m = [(n-1)/2]$. The notation $[x]$ implies taking the integer part of x.

$\underset{\sim}{g}$ is a binary vector with

$$g_k = \begin{cases} 1 \text{ if set } \{k\} \text{ is included in the design} \\ 0 \text{ otherwise} \end{cases}$$

It turns out that another measure of connectedness, conceptually rather different from the notion of analysis of variance efficiency, is mathematically identical to efficiency and also can be easily

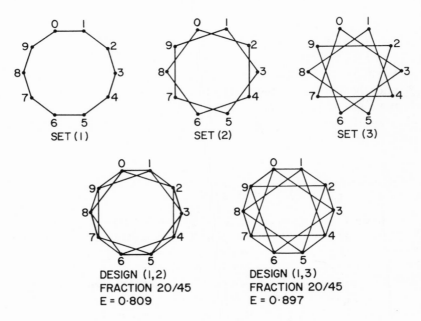

Figure 1. Three cyclic sets and two designs for $n = 10$.

calculated for any design, including random and other unbalanced structures. This is the mean electrical resistance of the graph of the design, with the edges considered as conductors of unit resistance. This index was introduced by Johnson and Van Dyk [11], who also showed that efficiency and mean resistance are mathematically identical. Since their work has not yet been published, it is perhaps worth indicating here how mean resistance can be conveniently calculated, although the theory behind the method will not be repeated.

1. Construct a matrix G with $g_{ij} = -1$ if the edge connecting i and j is present, and zero otherwise. The diagonal of G has elements,

$$g_{ii} = \sum_{j \neq i}^{n} |g_{ij}|$$

2. Invert the submatrix consisting of the first $n - 1$ rows and columns of G. Augment this with a row and column of zeros. Call the result H.
3. The mean resistance of the network is

$$\overline{R} = \sum_{i,j}^{n} (h_{ii} + h_{jj} - 2h_{ij})/n(n - 1)$$

Mean resistance has considerable intuitive appeal as a measure of connectedness, since its magnitude decreases when paths are either shorter or more numerous. As pointed out by Spence and Domoney [19], the number of triangles in the graph of any design is very closely, though not perfectly, related to both of the above measures.

THE RELATION BETWEEN CONNECTEDNESS
AND RECOVERY

The following simple experiment was performed to demonstrate the dependence of recovery on the connectedness of the design. Forty points, arranged in the fashion of a Saint Andrew's cross and scaled so that the maximum interpoint distance was two units, formed the basis for constructing the incomplete data sets used in the scalings. (See the first diagram in Figure 2.) The size of this configuration was comparable to those employed in other experiments [18, 19]; thus, the level of error used in this demonstration can be immediately related to the previous studies. The dissimilarities were computed by using an error model that has been extensively discussed by Ramsay [14]. A low-moderate level of error was used, with standard deviation equal to 0.1, and each dissimilarity was computed as follows:

$$\delta_{ij} = [\sum_{a=1}^{m} (x_{ia}^e - x_{ja}^e)^2]^{\frac{1}{2}}$$

with $x_{ia}^e = x_{ia} + N(0,0.01)$ and $m = 2$. Only one error level was used, since the Spence-Domoney study did not reveal any substantial interactions involving designs and the level of error. A two-dimensional true configuration was chosen, so that a visual impression of the quality of recovery could be obtained, in addition to the various possible statistical measures.

Twenty-seven different cyclic designs were used to create incomplete data sets based on the same dissimilarity matrix; in all designs a 320/780 fraction, or 41%, of the data was retained. Hence, the "degrees of freedom ratio" (number of pairs/$(m(n-1) - m(m-1)/2)$ [19, 22]) for all designs was 4.16, thus marginally exceeding the recommended minimum of about 3.5 proposed [19]. Each of the cyclic designs was composed of eight cyclic sets; these designs are listed in the first column of Table 1. This table clearly shows the strong relation between the number of triangles in a design and its McKeon-David efficiency. Mean resistance is not reported, since it is directly proportional to efficiency. It also can be seen that it is possible to create designs that differ slightly in efficiency but have the same number of triangles—witness the two zero triangle designs.

The other columns in Table 1 contain the statistics that summarize the results of the twenty-seven scalings, which were performed using TORSCA-9 with two iterations in the quasi-nonmetric phase and fifty in the fully nonmetric phase [21]. Shown also are the correlations between the known and recovered distances (r), Kruskal's stress formula one, and EFIT. This latter index is a measure of discrepancy between a recovered configuration that has been translated, dilated, and orthogonally rotated to a least squares best fit to a specified target [17]. In this case, the target was the Saint Andrew's cross configuration used to generate the data. The target and the results obtained from five of the designs, equally spaced in terms of the number of triangles, are shown in Figure 2. At first sight, it may seem strange that two of the solutions are much "smaller" than the others. The reason is that portions of

TABLE 1
Connectedness and Recovery Measures
for Various Designs

Design	Number of Triangles	E	r	EFIT	Stress
1, 3, 5, 7, 9, 11, 15, 17	0	0.975	0.982	0.624	0.091
1, 3, 5, 7, 9, 11, 13, 15	0	0.974	0.980	0.649	0.090
6, 7, 8, 9, 10, 11, 12, 13	120	0.974	0.978	0.704	0.080
5, 7, 8, 9, 10, 11, 12, 13	200	0.972	0.982	0.607	0.078
5, 6, 7, 8, 9, 10, 11, 12	240	0.972	0.983	0.606	0.074
4, 7, 8, 9, 10, 11, 12, 13	280	0.970	0.977	0.700	0.080
1, 3, 5, 7, 9, 11, 16, 19	320	0.970	0.986	0.545	0.100
1, 2, 5, 7, 9, 11, 15, 17	360	0.969	0.981	0.680	0.098
4, 5, 6, 7, 8, 9, 10, 11	400	0.967	0.967	0.871	0.086
4, 5, 6, 7, 8, 9, 10, 12	440	0.967	0.983	0.623	0.077
4, 5, 6, 7, 8, 9, 10, 13	480	0.966	0.982	0.625	0.082
1, 2, 3, 4, 7, 9, 13, 19	520	0.965	0.974	0.734	0.103
1, 3, 6, 8, 11, 13, 16, 17	560	0.964	0.985	0.568	0.098
1, 2, 4, 5, 8, 10, 15, 16	600	0.963	0.984	0.609	0.103
1, 2, 3, 4, 5, 15, 16, 17	640	0.959	0.766	3.890	0.192
1, 2, 3, 4, 5, 6, 15, 16	680	0.957	0.750	3.975	0.183
1, 9, 10, 11, 12, 13, 14, 19	720	0.956	0.966	0.939	0.090
1, 2, 3, 4, 5, 9, 10, 11	760	0.949	0.981	0.641	0.095
1, 2, 14, 15, 16, 17, 18, 19	800	0.950	0.981	0.647	0.113
1, 2, 3, 4, 5, 6, 7, 15	840	0.942	0.724	4.044	0.185
1, 2, 3, 4, 5, 6, 7, 14	880	0.939	0.655	3.186	0.199
1, 2, 3, 4, 5, 6, 7, 13	920	0.936	0.964	0.918	0.107
1, 2, 3, 4, 16, 17, 18, 19	960	0.925	0.621	4.110	0.180
1, 2, 3, 4, 5, 6, 7, 11	1000	0.926	0.641	3.625	0.161
1, 2, 3, 4, 5, 6, 7, 10	1040	0.921	0.641	3.969	0.150
1, 2, 3, 4, 5, 6, 7, 9	1080	0.914	0.665	4.000	0.171
1, 2, 3, 4, 5, 6, 7, 8	1120	0.907	0.548	4.314	0.173

these solutions are on the "wrong" side(s) of the configuration, and, consequently, the FITA2B algorithm has had to shrink these configurations in order to optimize the least squares loss function.

Recovery is very good with all designs down to about the 600-triangle level, thereafter becoming quite erratic, with several solutions definitely unacceptable. This impression is supported by the high correlations between the recovery measures (*r* and EFIT) and the number of triangles and efficiency shown in Table 2. Poor recovery is also evidently associated with inflated stress values.

It is difficult to say why designs with low efficiencies perform more poorly. Obviously, if the starting configurations for the

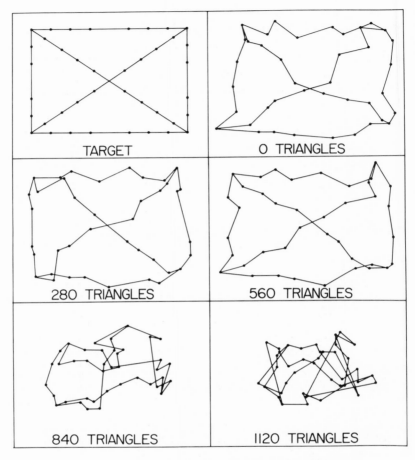

Figure 2. The target and five recovered configurations.

TABLE 2
Correlations of Connectedness and
Recovery Measures

	Number of Triangles	Efficiency	Recovery r	EFIT	Stress
Efficiency	-0.924	—			
Recovery r	-0.752	0.853	—		
EFIT	0.720	-0.783	-0.967	—	
Stress	0.720	-0.720	-0.906	0.942	—

algorithm were identical to the known configuration, then, in the nonmetric phase, the algorithm would probably move the configuration very little, *irrespective of the particular pattern of data retained*. Somehow designs with low efficiencies increase the susceptibility of the algorithm to being trapped in suboptimal positions, although it is clear from Table 1 that it is not invariably the case that lower-efficiency designs are incapable of obtaining good solutions. (See, for example, the 720-, 760-, and 800-triangle designs in Table 1.) A low-efficiency design is, nevertheless, more likely to result in unsatisfactory performance, whereas high-efficiency designs consistently yield good results. At an intuitive level, it seems reasonable that data collected according to a design with high connectedness should perhaps be less likely to prove difficult; unfortunately, no more precise explanation has as yet been proposed. Nevertheless, the empirical observation stands: high-efficiency designs—and they need not be maximum-efficiency designs—outperform low-efficiency designs.

CONSTRUCTING HIGH-EFFICIENCY DESIGNS

Here the advice of David [5] is most useful. Additionally, other optimal cyclic designs have been constructed and tabulated [10]; unfortunately, however, only for $n \leqslant 30$, so the practical usefulness of these tables for multidimensional-scaling purposes is limited. At least two other approaches can be mentioned. The first employs an algorithm that will search for a high-efficiency design (and generally finds an optimal design), whereas the second is a much simpler paper-and-pencil technique. Both provide adequate designs.

The Exchange Algorithm

This procedure is based on Mitchell's [13] D-optimal design program, DETMAX, which was developed for a similar purpose. In the following description, reference is made to the notation introduced earlier to describe cyclic designs:

1. Choose an arbitrary g_0 with q cyclic sets.
2. (a) Form g_i by adding a cyclic set such that the resulting $(q + 1)$-set design has maximal efficiency
   OR
   (b) Form g_i by deleting a cyclic set such that the resulting $(q - 1)$-set design has maximal efficiency.
3. Move to a q-set design by adding or deleting a set as necessary — always choosing the set to yield maximum efficiency.

4. Iterate 2 and 3 until convergence is obtained (guaranteed, at least to a local optimum).

This method can be modified by allowing *excursions* in step 2. These require the addition or subtraction of *more than one set* — in practice this yields only a very minor gain at the expense of increased computing time. Consequently, it is probably not desirable to increase the level of complexity in step 2. The initial choice of a or b in step 2 can be made randomly. An example of the operation of the algorithm is given in Table 3, for $n = 25$ and $r = 6$.

TABLE 3
A Small Example

														$E(q)$	$E(q + 1)$
$\underset{\sim}{g}_0$	=	[1	1	1	0	0	0	0	0	0	0	0	0]	0.654	
$\underset{\sim}{g}_9$	=	[1	1	1	0	0	0	0	0	1	0	0	0]		0.907
$\underset{\sim}{g}_{13}$	=	[0	1	1	0	0	0	0	0	1	0	0	0]	0.889	
$\underset{\sim}{g}_{16}$	=	[0	1	1	0	0	0	0	0	0	0	0	0]		0.709
$\underset{\sim}{g}_{26}$	=	[0	1	1	0	0	0	0	0	1	0	0	0]	0.889	
$\underset{\sim}{g}_{35}$	=	[0	1	1	0	0	0	0	0	1	1	0	0]		0.935
$\underset{\sim}{g}_{39}$	=	[0	1	0	0	0	0	0	0	1	1	0	0]	0.893	
$\underset{\sim}{g}_{42}$	=	[0	0	0	0	0	0	0	0	1	1	0	0]		0.763
$\underset{\sim}{g}_{52}$	=	[0	1	0	0	0	0	0	0	1	1	0	0]	0.893	
$\underset{\sim}{g}_{61}$	=	[0	1	1	0	0	0	0	0	1	1	0	0]		0.935
$\underset{\sim}{g}_{65}$	=	[0	1	0	0	0	0	0	0	1	1	0	0]	0.893	
$\underset{\sim}{g}_{69}$	=	[0	0	0	0	0	0	0	0	1	1	0	0]		0.763
$\underset{\sim}{g}_{79}$	=	[0	1	0	0	0	0	0	0	1	1	0	0]	0.893 (optimum)	

NOTE: Starting $\underset{\sim}{g}_0$ is worst possible; $n = 25$; $r = 6$.

The algorithm has been applied to 136 problems with known optimal solutions [10, Table A]. It found 124 optimal designs, one of which had higher efficiency than the tabulated design — Design A51 is in error. The 12 "failures" had mean efficiency only 0.0035 lower than the optimal designs, with a maximum discrepancy of 0.0160. All 136 designs were constructed on a CDC CYBER 73/14 in 50 seconds.

A Simpler Approach

Suppose that r replications of each stimulus are desired, thus yielding a $r/(n - 1)$ fraction; then, the required number of cyclic sets is $q = [(r + 1)/2]$. If r is odd and n is even, then the half set must be included in the design [5]. The vector $\underset{\sim}{g}$, which contains $p = [n/2]$ elements, can then be chosen so that the ones are equally spaced. For example:

$n = 20 \quad r = 7$
$\underset{\sim}{g} = [1\ 0\ 0\ 1\ 0\ 0\ 1\ 0\ 0\ 1]$
$n = 41 \quad r = 20$
$\underset{\sim}{g} = [1\ 0\ 1\ 0\ 1\ 0\ 1\ 0\ 1\ 0\ 1\ 0\ 1\ 0\ 1\ 0\ 1\ 0\ 1\ 0]$
$n = 30 \quad r = 20$
$\underset{\sim}{g} = [1\ 1\ 0\ 1\ 1\ 0\ 1\ 1\ 0\ 1\ 1\ 0\ 1\ 1\ 0]$

In the last example, equal spacing is not possible because of the number of sets required; consequently, approximately equal spacing of small groups has been used. This works quite well and, in any case, will only be necessary when more than 50% of the data is being collected; thus, a good solution could probably be obtained with almost any design.

This rather simple approach will always produce a design that is quite close to an optimal design. In some instances, it may even be optimal. Green, Kehoe, and Reynolds [9] have proposed an essentially equivalent procedure in connection with their work on interactive ordering.

AN EMPIRICAL APPLICATION

In order to allow informal comparison with the Young and Cliff [23] interactive scaling study, their stimulus set of forty-two verbs of human locomotion was used; the verb PLOD was included twice as a reliability check, making forty-three stimuli in total. The high-efficiency cyclic design employed was composed of twelve cyclic sets; hence, 512 of the possible 903 pairwise judgments were collected. This is somewhat more than the average number of judgments (from 226 to 393) collected in the Young-Cliff experiment; in fact, as it will be shown, good results could have been obtained with fewer judgments, but the conservative rule suggested by Spence and Domoney [19] was used: one should obtain about three or four times as many judgments as there are coordinates in the expected solution. On the basis of the Young-Cliff results, the recovered configuration was expected to be at least three dimensional.

The experimental instructions to the nine subjects were the same as in the Young-Cliff study. The subjects sorted computer cards, with the stimulus pairs printed along the top, into nine categories. Category 1 was labeled "highly similar" and category 9, "highly dissimilar." The subjects took from 1.2 to 2.1 hours to complete the task; the average was 1.6 hours. This is about the same time as in the Young-Cliff study, although their subjects made fewer judgments.

Using TORSCA-9 [21], solutions were obtained in five through one dimensions. All solutions were stable, with no suggestion of degeneracies. An examination of the stress values suggested that most subjects needed two or three dimensions to fit their judgment data. The two-dimensional solution of a typical subject is shown in Figure 3. This plot is very similar to those published by Young

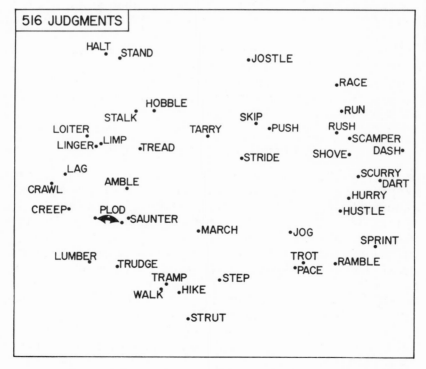

Figure 3. The two-dimensional solution of a typical subject, with a 57% fraction.

and Cliff [23]. Some indication of the reliability can be gained by observing the closeness of the two PLODs.

We can get some idea of whether or not an acceptable solution could have been obtained if the subject had given fewer than 516 judgments. Two cyclic sets were discarded and the remaining 430 judgments were submitted to TORSCA-9; two more sets were discarded, leaving 344 judgments; and finally the multidimensional scaling algorithm was applied to the 258 judgments left after removing two more cyclic sets. At each stage, the cyclic sets removed were chosen so as to leave that set that yielded the highest-efficiency design. The obtained configurations, rotated to congruence, are shown in Figures 4, 5, and 6.

It is easy to see that not much has been lost by reducing the number of judgments. Though an interpretation of the results of the scalings will not be attempted here, it is probably safe to say that conclusions based on any of Figures 3 to 6 would not differ

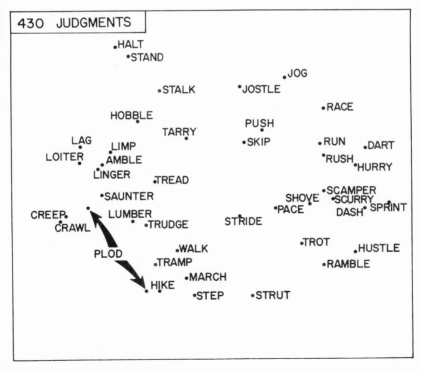

Figure 4. The two-dimensional solution of a typical subject, with a 47% fraction.

substantially. There is a small problem in Figures 4 and 5: the PLODs seem to be rather far apart. This is probably an indication of a minor local minimum problem; and in practice this might alert a user to restart, adjust algorithm parameters, or increase the number of iterations. In fact, the final solutions in these cases were subsequently improved by such a "fine-tuning" process, but the original results are presented to illustrate this phenomenon.

The five-dimensional solutions, which were obtained from the nine subjects' judgment data were used as input to the INDSCAL algorithm [3]. This was done to demonstrate that it is possible to examine individual differences, if desired, and further to suggest a useful method of obtaining an average space for the subjects. The two-dimensional group stimulus space (unrotated) from INDSCAL is presented in Figure 7. This configuration again is seen to be strikingly similar to the average spaces obtained by Young and Cliff

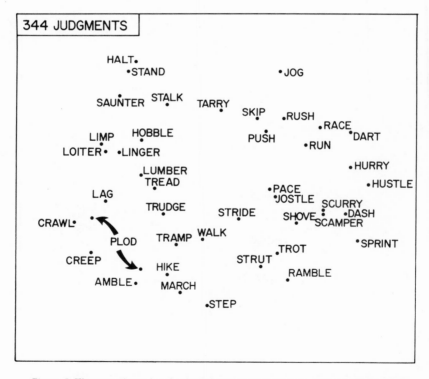

Figure 5. The two-dimensional solution of a typical subject, with a 38% fraction.

[23] and especially similar to the maximum quality group space (their Figure 6).

CONCLUSIONS

Probably the most important decision to be made is the size of the fraction to be collected. It would seem that it is dangerous to employ fractions with degrees of freedom ratios less than about three, independent of the design chosen [19]. A good rule of thumb might be to guess the likely dimensionality (erring on the conservative side) and then multiply this number by $3n$ to yield the absolute minimum number of judgments necessary. Thus, for forty points in three dimensions at least $3 \times 40 \times 3 = 360$ judgments will be necessary—this represents a 46% fraction of the data. In general, the minimum fraction for n points in m dimensions will be

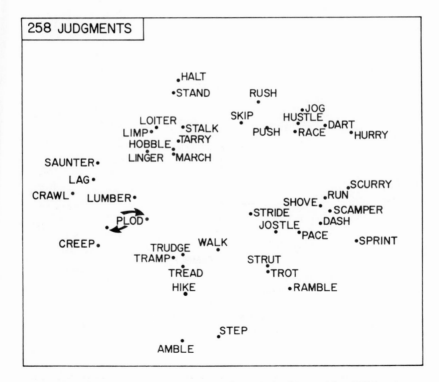

Figure 6. The two-dimensional solution of a typical subject, with a 29% fraction.

$$F_{min} = 3nm/(n(n-1)/2)$$
$$= 6m/(n-1)$$

Hence, $r = 6m$ replications of each stimulus will be required. Clearly, as n increases, the minimum fraction necessary will become smaller. Likewise, spaces that are expected to be of high dimensionality will necessitate the use of larger fractions.

Even with a conservative fraction, a bad design can yield unacceptable results. It is the recommendation of this author that high-efficiency cyclic designs be considered in view of their proven performance, ease of construction, and balance. However, other designs can also be used. Even random designs will usually perform satisfactorily. Also, novel designs created by the experimenter can be compared to high-efficiency cyclic designs in order to determine their acceptability; a comparison of the mean resistance of the novel design to the mean resistance of a high-efficiency cyclic de-

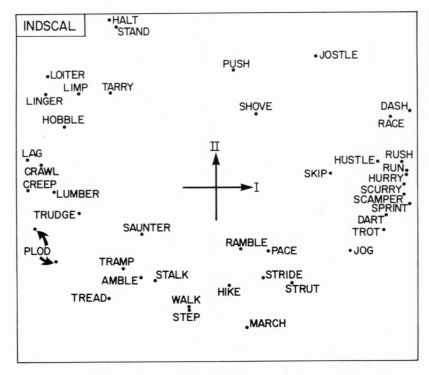

Figure 7. The INDSCAL two-dimensional group stimulus space.

sign can be used to settle the question of whether the new design is any good.

Another, rather different, approach to the problem has recently been proposed by Graef and Spence [8]; they showed that the large distances are crucially important in multidimensional scaling. If the large distances are lost, recovery deteriorates dramatically. This is not the case when the small or the medium distances are missing. Consequently, if an investigator has prior knowledge of the likely magnitudes of the dissimilarities, then the small or medium dissimilarity comparisons can be eliminated when the incomplete set of pairs to be presented to the subject is chosen. This prior information is surprisingly easy to come by, and the reader is referred to the article by Graef and Spence [8] for further discussion, including the relationship of these "pseudo designs" to the more formal designs examined in this chapter.

REFERENCES

1. Baker, R. F., and F. W. Young, 1974. *Interactive Scaling with Individual Subjects II: An Improved System.* Chapel Hill, N.C.: University of North Carolina. L. L. Thurstone Psychometric Laboratory Report #132.

2. Boorman, S. A., and P. Arabie, 1972. "Structural Measures and the Method of Sorting." In R. N. Shepard, A. K. Romney, and S. B. Nerlove (eds.), *Multidimensional Scaling: Theory and Applications in the Behavioral Sciences* (vol. 1), pp. 225-49. New York: Seminar Press.

3. Carroll, J. D., and J. J. Chang, 1970. "Analysis of Individual Differences in Multidimensional Scaling via an N-Way Generalization of 'Eckart-Young' Decomposition," *Psychometrika*, 35, 283-319.

4. Clatworthy, W. H., 1973. *Tables of Two-Associate Class Partially Balanced Designs.* Washington, D.C.: National Bureau of Standards Applied Mathematics Series #63.

5. David, H. A., 1963. "The Structure of Cyclic Paired-Comparison Designs," *Journal of the Australian Mathematical Society*, 3, 117-27.

6. Fillenbaum, S., and A. Rapoport, 1972. "An Experimental Study of Semantic Structures." In A. K. Romney, R. N. Shepard, and S. B. Nerlove (eds.), *Multidimensional Scaling: Theory and Applications in the Behavioral Sciences* (vol. 2), pp. 93-131. New York: Seminar Press.

7. Girard, R. A., and N. Cliff, 1976. "A Monte Carlo Evaluation of Interactive Multidimensional Scaling," *Psychometrika*, 41, 43-64.

8. Graef, J., and I. Spence, 1979. "Using Distance Information in the Design of Large Multidimensional Scaling Experiments," *Psychological Bulletin*, 86, 60-66.

9. Green, R. S., J. F. Kehoe, and T. J. Reynolds, 1975. "An Evaluation of Interactive Ordering." Unpublished manuscript, University of Southern California, Los Angeles.

10. John, J. A., F. W. Wolock, and H. A. David, 1972. *Cyclic Designs.* Washington, D.C.: National Bureau of Standards Applied Mathematics Series #62.

11. Johnson, R. M., and G. J. Van Dyk, 1975. "A Resistance Analogy for Efficiency of

Paired Comparison Designs," a paper presented at the Annual Meetings of the Psychometric Society, University of Iowa, Iowa City, Iowa, 1975.

12. McKeon, J. J., 1960. *Some Cyclical Incomplete Paired Comparison Designs.* Chapel Hill, N.C.: University of North Carolina, L. L. Thurstone Psychometric Laboratory Report #24.

13. Mitchell, T. J., 1974. "An Algorithm for the Construction of 'D-Optimal' Experimental Designs," *Technometrics*, 16, 203-10.

14. Ramsay, J. O., 1969. "Some Statistical Considerations in Multidimensional Scaling," *Psychometrika*, 34, 167-82.

15. Rao, V. R., and R. Katz, 1971. "Alternative Multidimensional Scaling Methods for Large Stimulus Sets," *Journal of Marketing Research*, 8, 488-94.

16. Rosenberg, S., and A. Sedlak, 1972 "Structural Representations of Perceived Personality Trait Relationships." In A. K. Romney, R. N. Shepard, and S. B. Nerlove (eds.), *Multidimensional Scaling: Theory and Applications in the Behavioral Sciences* (vol. 2), pp. 133-62. New York: Seminar Press.

17. Schönemann, P. M., and R. M. Carroll, 1970. "Fitting One Matrix to Another under Choice of a Central Dilation and a Rigid Motion," *Psychometrika*, 35, 245-56.

18. Spence, I., 1972. "A Monte Carlo Evaluation of Three Nonmetric Multidimensional Scaling Algorithms," *Psychometrika*, 37, 461-86.

19. Spence, I., and D. W. Domoney, 1974. "Single Subject Incomplete Designs for Nonmetric Multidimensional Scaling," *Psychometrika*, 39, 469-90.

20. Torgerson, W. S., 1958. *Theory and Methods of Scaling.* New York: John Wiley and Sons.

21. Young, F. W., 1968. *A FORTRAN IV Program for Nonmetric Multidimensional Scaling.* Chapel Hill, N.C.: University of North Carolina, L. L. Thurstone Psychometric Laboratory Report #56.

22. Young, F. W., 1970. "Nonmetric Multidimensional Scaling: Recovery of Metric Information," *Psychometrika*, 35, 455-73.

23. Young, F. W., and N. Cliff, 1972. "Interactive Scaling with Individual Subjects," *Psychometrika*, 37, 385-415.

Chapter 1.4 Sampling Designs and Recovering Cognitive Representations of an Urban Area

Aron N. Spector and *Victoria L. Rivizzigno*

The general aim of this paper is to examine multidimensional scaling as a means of recovering information concerning how people cognize the relative locations of places in the urban environment. This examination leads to a more specific aim of determining how best to collect and analyze large sets of scaling data.

An experiment to uncover cognitive information about an urban environment focused on Columbus, Ohio, where individuals were asked to evaluate the relative spatial separation of a sample of locations drawn from the Columbus metropolitan area. These distance judgments were then subjected to multidimensional scaling analysis.

In this analysis, only a subset of the distance relationships among these locations was inserted in the scaling algorithm. Since sampled estimates were presumed to contain error, these distances were also perturbed with a common error model. The object of this presentation is the examination of how robust this algorithm behaves with different modes of organizing the sampling of distance data under different magnitudes of error.

The research for this paper was supported in part by a grant (GS-37969) from the National Science Foundation for the project Cognitive Configurations of a City, awarded to R. G. Golledge and J. N. Rayner, Department of Geography, Ohio State University.

THE LOCATION EXPERIMENT

Location Stimuli

The locations used in the experiment were chosen on the basis of general population familiarity and on the principle of maximum spatial dispersion. These locations are shown in Figure 1 and are identified in Table 1. The procedure for choosing the forty-nine locations is fully described in Rivizzigno [13].

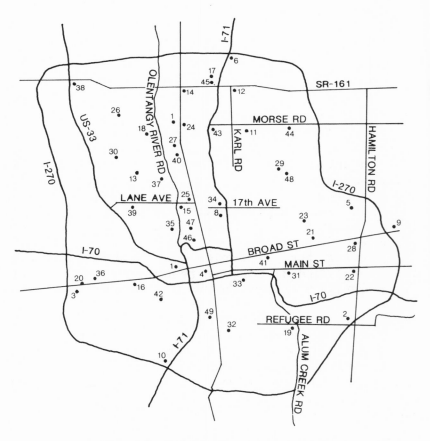

Figure 1. Spatial distribution of sample locations.

TABLE 1
Sample Locations

Location Number	Location Name
1	Graceland Shopping Center
2	Eastland Shopping Center
3	Westland Shopping Center
4	Lazarus, downtown
5	Port Columbus
6	I-71 north and I-270 intersection
7	Veterans' Memorial Auditorium
8	Ohio State Fairgrounds Coliseum
9	Western Electric, east industrial plant
10	I-71 south and I-270 intersection
11	Northland Shopping Center
12	Karl Rd. and S.R. 161 intersection
13	Kingdale Shopping Center
14	N. High St. and S.R. 161 intersection
15	OSU Football Stadium
16	Columbus State Hospital
17	Anheuser Busch Brewery
18	Henderson Rd. and Reed Rd. intersection
19	Refugee Rd. and Alum Creek Dr. intersection
20	W. Broad St. and Georgesville Rd. intersection
21	Town and Country Shopping Center
22	Hamilton Rd. and E. Main intersection
23	Ohio Dominican College
24	Morse Rd. and N. High St. intersection
25	Lane Ave. and N. High St. intersection
26	Don Scott Air Field
27	Henderson Rd. and N. High St. intersection
28	E. Broad St. and Hamilton Rd. intersection
29	Northern Lights Shopping Center
30	Golden Bear Shopping Center
31	Capital University
32	Schottenstein's South
33	Grant Hospital
34	Ohio Historical Society
35	Jai Lai Restaurant
36	Great Western Shopping Center
37	Riverside Hospital
38	Dublin Rd. and S.R. 161 intersection
39	Lane Avenue Shopping Center
40	Whetstone Park
41	Franklin Park
42	Central Point Shopping Center
43	I-71 north and Morse Rd. intersection
44	Cleveland Ave. and Morse Rd. intersection
45	French Market
46	Thurber Club Apartments
47	Battelle Memorial Institute
48	Schottenstein's North
49	Southview Park

Type of Data Collected

The distance judgments the subjects were asked to make were in the form of paired comparisons. Subjects were asked to use a scale from one to nine and indicate whether a pair of locations was very close together (scale value of one), very far apart (scale value of nine), or somewhere in between (scale values two through eight). [See 13 for details regarding the experimental procedures.]

The number of paired comparisons used in the location experiment was restricted because of the subjects' limited capabilities of completing large tasks. To obtain full information concerning all comparisons between pairs of the test locations, by sampling each pair once, $n(n - 1)/2$ comparisons would have been required (in our case, 1,176). It is obvious that such a large number of comparisons provides a rather cumbersome weight for any subject. Thus, (1) accuracy may suffer from the subjects' weariness and boredom, and (2) the subject may become lost or disoriented when faced with such an onerous task. Furthermore, because of the high degree of redundancy inherent in paired comparison data, it has been postulated that only a subsample of comparisons is necessary for the recovery of metric information, even when large amounts of sampling error are present.

A number of measures have, therefore, been put forward to determine: (1) whether the number of comparisons chosen is sufficient for the recovery of metric information and (2) whether the pattern of comparisons (i.e., which subset of the population of all possible comparisons) is most reliable in recovering metric information, given a fixed number of comparisons. In this second regard, an emphasis has been placed on the determination of an appropriate sampling design. A number of sampling designs have been examined and put forward by Torgerson [17] and Spence and Domoney [16], among others. Of prime interest in the evaluation of the above criteria is the comparison of the sampling design used in this study with other more conventional designs.

SAMPLING DESIGN FOR THE
LOCATION EXPERIMENT

Once it was decided to use an incomplete matrix design, a set of criteria was developed to determine which of the possible 1,176 pairs of locations were to be used. This was accomplished in light of another procedure that was used to lighten the burden of the subjects. The chosen pairs of locations were to be presented in five different subsamples, instead of all at once (Table 2).

In all, 437 comparisons of different pairs of places (37.2% of all possible pairs) were collected from each subject. To check for continuity, overlaps were included so that some comparisons were made more than once. The criteria used to determine which comparison was selected were: (1) all paired comparisons for a subset of locations would be obtained during each sampling period (after a suggestion from Torgerson [17, pp. 191-92]) (2) each subsample had the characteristic that there was a relatively uniform distribution of real distances between the places chosen as cues; and (3) anchor points representing the places nearest to each other and farthest from each other were used in each set.

This last stipulation was included because of a tendency for subjects to "rescale" the distances between points relative to extremes in distances during each occasion when a subsample was taken. During our preliminary investigation, this led to marked discontinuities [13]. The pattern of comparisons is shown in Figure 2 as a set of five overlapping steps with a set of vertical lines (in columns 4, 6, 7 and 10). The figure indicates the use of a combination of the sampling methods described in Figures 5C and 5D of Torgerson [17, pp. 192-93].

ALTERNATIVE SAMPLING DESIGNS

The selection of the location experiment sampling design preceded an examination by Spence and his colleagues of a number of alternative sampling designs used in the collection of incomplete paired comparison data [3, 16]. A number of these are well known in the area of analysis of variance and include partially balanced incomplete designs (PBID).

An advantage of this set of designs is indicated by the sufficiency criteria outlined by Kendall [8] as minimum requirements for any given design. First, every object should be used an equal number of times within a design. Second, it should be impossible to separate objects into two equally sized subsets so that no comparisons are made between any object in one and any object in the other. Both criteria are met using the PBID design. In contrast, only the second criterion is met by the location experiment design.

Partially balanced incomplete designs, in general, have the following characteristics: (1) There are n points and b comparisons (referred to as "blocks" in the sampling design literature). (2) Each

TABLE 2
Sample Locations Subset

Location Number	Location Name
	Subset 1
1	Graceland Shopping Center
2	Eastland Shopping Center
3	Westland Shopping Center
4	Lazarus, downtown
5	Port Columbus
6	I-71 north and I-270 intersection
7	Veterans' Memorial Auditorium
8	Ohio State Fairgrounds Coliseum
9	Western Electric, east industrial plant
10	I-71 south and I-270 intersection
11	Northland Shopping Center
12	Karl Rd. and S.R. 161 intersection
13	Kingdale Shopping Center
14	N. High St. and S.R. 161 intersection
15	OSU Football Stadium
	Subset 2
14	N. High St. and S.R. 161 intersection
15	OSU Football Stadium
16	Columbus State Hospital
4	Lazarus, downtown
17	Anheuser Busch Brewery
6	I-71 north and I-270 intersection
7	Veterans' Memorial Auditorium
18	Henderson Rd. and Reed Rd. intersection
19	Refugee Rd. and Alum Creek Dr. intersection
20	W. Broad St. and Georgesville Rd. intersection
21	Town and Country Shopping Center
22	Hamilton Rd. and E. Main intersection
23	Ohio Dominican College
24	Morse Rd. and N. High St. intersection
	Subset 3
23	Ohio Dominican College
24	Morse Rd. and N. High St. intersection
25	Lane Ave. and N. High St. intersection
4	Lazarus, downtown
26	Don Scott Air Field
6	I-71 north and I-270 intersection
7	Veterans' Memorial Auditorium
27	Henderson Rd. and N. High St. intersection
28	E. Broad St. and Hamilton Rd. intersection
29	Northern Lights Shopping Center
30	Golden Bear Shopping Center
31	Capital University
32	Schottenstein's South

TABLE 2 (Continued)

Location Number	Location Name
	Subset 4
31	Capital University
32	Schottenstein's South
33	Grant Hospital
4	Lazarus, downtown
34	Ohio Historical Society
6	I-71 north and I-270 intersection
7	Veterans' Memorial Auditorium
35	Jai Lai Restaurant
36	Great Western Shopping Center
37	Riverside Hospital
38	Dublin Rd. and S.R. 161 intersection
39	Lane Avenue Shopping Center
40	Whetstone Park
	Subset 5
39	Lane Avenue Shopping Center
40	Whetstone Park
41	Franklin Park
4	Lazarus, downtown
42	Central Point Shopping Center
6	I-71 north and I-270 intersection
7	Veterans' Memorial Auditorium
43	I-71 north and Morse Rd. intersection
44	Cleveland Ave. and Morse Rd. intersection
45	French Market
46	Thurber Club Apartments
47	Battelle Memorial Institute
48	Schottenstein's North
49	Southview Park

point occurs r times so that $nr = 2b$ (i.e., there are a total of $2b$ comparisons). (3) Each pair of points occurs $L1$ or $L2$ times. (These constants are decided on by the experimenter.) (4) Each object is compared to $n1$ other objects and indirectly compared to $n2$ other objects. (Note that an indirect comparison is defined when both objects i and j are compared to a third object k.) (5) The design is such that, given two objects connected only indirectly through i comparisons, the number of objects connected to the first object via j comparisons and the second object via k comparisons is always equal to a constant (p). (Note that a connection is defined when a network of comparisons can be defined linking two objects. Thus, if i is compared to j, j to k, k to l, and l to m, then i is connected to m via four comparisons [1].)

```
 2  x
 3  xx
 4  xxx
 5  xxxx
 6  xxxxx
 7  xxxxxx
 8  xxxxxxx
 9  xxxxxxxx
1Ø  xxxxxxxxx
11  xxxxxxxxxx
12  xxxxxxxxxxx
13  xxxxxxxxxxxx
14  xxxxxxxxxxxxx
15  xxxxxxxxxxxxxx
16     x xx  x  xxx
17     x xx  x  xxxx
18     x xx  x  xxxxx
19     x xx  x  xxxxxx
2Ø     x xx  x  xxxxxxx
21     x xx  x  xxxxxxxx
22     x xx  x  xxxxxxxxx
23     x xx  x  xxxxxxxxxx
24     x xx  x  xxxxxxxxxxx
25     x xx  x          xxx
26     x xx  x          xxxx
27     x xx  x          xxxxx
28     x xx  x          xxxxxx
29     x xx  x          xxxxxxx
3Ø     x xx  x          xxxxxxxx
31     x xx  x          xxxxxxxxx
32     x xx  x          xxxxxxxxxx
33     x xx  x                  xxx
34     x xx  x                  xxxx
35     x xx  x                  xxxxx
36     x xx  x                  xxxxxx
37     x xx  x                  xxxxxxx
38     x xx  x                  xxxxxxxx
39     x xx  x                  xxxxxxxxx
4Ø     x xx  x                  xxxxxxxxxx
41     x xx  x                          xxx
42     x xx  x                          xxxx
43     x xx  x                          xxxxx
44     x xx  x                          xxxxxx
45     x xx  x                          xxxxxxx
46     x xx  x                          xxxxxxxx
47     x xx  x                          xxxxxxxxx
48     x xx  x                          xxxxxxxxxx
49     x xx  x                          xxxxxxxxxxx
```

x denotes an included paired comparison

Figure 2. Our design.

Four such designs have been defined, including the commonly used cyclic design. Most of these designs can be easily constructed. To construct a cyclic design, rows and columns are numbered in an incidence matrix (N) of possible comparisons. Points are numbered 1 to n. If a comparison is made between point i and point j, then $N(i,j) = 0$. Each "cycle" defines different comparisons beginning in any arbitrary row(s) and column 1.

Comparisons used in each cycle thus are defined by:

$$N(s + k, k), k = 1, 2 \ldots r$$
$$N(r + k, k - r), k = r + 1 \ldots n$$

where n defines the number of points, s defines the row in which the cycle begins, and $r = n - s$. ($s = n/2 + 1$ when n is odd; $s = n/2$ when n is even.)

An arbitrary number of such cycles can be used, depending on the proportion of information desired. Figure 3 shows an example of such a design.

Another attractive alternative that does not guarantee either of Kendall's criteria is the selection of a random design, where n comparisons are arbitrarily assigned. Given a moderate number of inclusions, it is rather improbable that Kendall's second criterion will not be met. (See Figure 4.)

Recently, Graef and Spence [3] suggested using designs that utilize aspects of the data as selection criteria. In particular, they have found that deleting small and medium distances leads to greater recovery of information using MDS under conditions of high error. This, in effect, defines a type of stratified sampling procedure. (See Figure 5.)

METHODOLOGY

In order to examine the relative adequacy of our location experiment design in terms of metric recovery, our design was compared to cyclic and random designs and to a design utilizing only largest distances. Since we collected 437 comparisons, the 437 largest distances and 437 random distances were used in the alternative designs. A few more comparisons (441) were used in the cyclic design, since this constituted nine full cycles.

Two sets of criteria were used to determine the relative adequacy of the designs. First, a number of measures proposed within the literature were examined and compared among these sample designs.

```
 2  x
 3    x
 4      x
 5  x     x
 6  x   x
 7      x   x
 8  x   x   x
 9    x   x   x
10      x   x   x
11  x   x   x   x
12    x   x   x   x
13      x   x   x   x
14  x   x   x   x   x
15    x   x   x   x   x
16  x x   x   x   x   x
17    x x   x   x   x   x
18  x x x   x   x   x   x
19    x x x   x   x   x   x
20      x x x   x   x   x   x
21  x   x x x   x   x   x   x
22    x   x x x   x   x   x   x
23      x   x x x   x   x   x   x
24  x   x   x x x   x   x   x   x
25    x   x   x x x   x   x   x   x
26      x   x   x x x   x   x   x   x
27  x   x   x   x x x   x   x   x   x
28    x   x   x   x x x   x   x   x   x
29      x   x   x   x x x   x   x   x   x
30  x   x   x   x   x x x   x   x   x   x
31    x   x   x   x   x x x   x   x   x   x
32      x   x   x   x   x x x   x   x   x   x
33  x   x   x   x   x   x x x   x   x   x   x
34    x   x   x   x   x   x x x   x   x   x   x
35  x x   x   x   x   x   x x x   x   x   x   x
36    x x   x   x   x   x   x x x   x   x   x   x
37  x x x   x   x   x   x   x x x   x   x   x   x
38    x x x   x   x   x   x   x x x   x   x   x   x
39      x x x   x   x   x   x   x x x   x   x   x   x
40  x   x x x   x   x   x   x   x x x   x   x   x   x
41    x   x x x   x   x   x   x   x x x   x   x   x   x
42      x   x x x   x   x   x   x   x x x   x   x   x   x
43  x   x   x x x   x   x   x   x   x x x   x   x   x   x
44    x   x   x x x   x   x   x   x   x x x   x   x   x   x
45      x   x   x x x   x   x   x   x   x x x   x   x   x   x
46  x   x   x   x x x   x   x   x   x   x x x   x   x   x   x
47    x   x   x   x x x   x   x   x   x   x x x   x   x   x   x   x
48      x   x   x   x x x   x   x   x   x   x x x   x   x   x   x   x
49  x   x   x   x   x x x   x   x   x   x   x x x   x   x   x   x   x
```

x denotes an included paired comparison

Figure 3. The cyclic design.

```
 2  x
 3
 4   xx
 5  x   x
 6      x
 7      xx
 8  x   x   x
 9   xx x
10
11       x      x
12  x       x
13   xx xx x
14  x    x xxx xxx
15       xx   x xxx
16  x    xxx   x xx x
17   x   xx   xx
18  x x   x x    x x
19       xx      x x x x
20  xx   xxxx x xx x    xx
21   x x   x    x      x x
22  xxx  x x         xx x   x
23   x  xx xxxxx xx x    xx
24        xx   x xx    x
25     x   x   x    x  xxxxx
26  xx x               x   x x
27   xx x x    xx x   xx     xxx
28  x  x       x     x x    xx
29       x x    xx     x   x xx x
30   x xx x xxxxxx   x   xx   xx
31  x x    x x   x          x   x    xx
32           x   x       x    x    x x
33   x    x  xx x        x    xxx x     x
34  xx xx xx x    xx x xx x    x
35   xx      xxx   x xxxx x x      xxxx
36  xxx   xx x    x   x xxx    x       x
37   x  x xxxx x     xxx    xx    x x xx x
38  xxxx x   x  x x    x   x    x x    x  x
39   xx x  xxxx     x    xx xxx    x   x
40  x   x x    x    x       x  xxxx       x
41  xx   x xx x xx x x      xx    x x x    x
42            x x xx x x xx xx  xxxxxxx      x
43   x  x    xxx   x   x x x   x    x    xx xx
44    x  xx  x     xx x   x   x    x
45   x  xx xxx x    x     x   x    x  xx  x xx  x
46  x       x   xx   x  x  xx    xx xx   x     xx
47  xxx   xxxxx     x   x     x xx   xx     x xx
48   xxx x x  x      xx      xx    x x x
49  x   xx  xxx   x  x x   x xxx xx    xxxxx      xx
```

x denotes an included paired comparison

Figure 4. Design based on random selection of comparisons.

```
 2  x
 3  xx
 4
 5  x x
 6   xxxx
 7          x
 8   x     x
 9  x xx xxx
1Ø  xx    xx xx
11   xxx    x xx
12   xxxx x xx
13   x   x    xx
14   xxxx x xx
15   x        xx
16  xx   xx   x xx x
17   xxxx x xx      x
18   xxxx x xx      x
19  x x   x     xxxxx xx
2Ø  xx   xx   x xx x   xxx
21  x x   x    xxxxx xxx x
22  x x   x    xxxxxxxxx x
23    x   x    x    x  xx x
24   xx x    xx        xxxx
25   x       xx           x
26   xxxx x xx      x   xxxxx
27   xx x    xx         xxxx
28  x x   x    xxxxxxxxx x    x xx
29   xx      xx        xx
3Ø  x   xx   xx        x xxx      x
31  x x   x     xxxx   xx x    x x.x   x
32  x     x   x xxxx   xx      x xx xx
33        x   x xx x   xx         x
34   xx   x   xx           x
35   x   xx   x  x      x      x      x
36  xx   xx   x xx x   xxx xxxx xxxx x
37   x   x    xx       x xx      x       x
38   xxxx xxxxx    xx   xxxxx x   xx xxxxxx
39   x   xx   xx       x xx      x
4Ø  xx x    xx        xxxx      x    x
41  x x   x    xx x   xx x      x    x         x
42  xx   xx   x xx x   xx   xx x x xx      x
43   xx      xx       x  xxxx      x  xx    x      x
44   xx      xx       x  xx x      x x    xx  x    x
45   xxxx x xx       x  xxxx      x   xxx xx      xx
46   x   x   x   x x   x               x        x
47   x   x   x       x     x      x         x
48   xx      x        xx        x    x x   x
49  x    xx   x xx x   xx      x xxx x      x       xxx
```

x denotes an included paired comparison

Figure 5. Design based on the 437 largest distances.

Second, for our sample locations, different sizes and types of randomized error were added to the distance relationships defined by these points. A MDS solution of relative locations was generated and compared to the original locations as an indication of metric determinacy. Although each of these two sets of criteria are important, *the final and all-important criterion must remain the recovery of a set of relative locations that match the model of sample locations as closely as possible.*

MEASURES RELATED TO THE ADEQUACY OF A SAMPLING DESIGN

Determining the Adequate Number of Points

According to Young [19], recovery of distance and locational information is related to the minimum number of distance estimates required in establishing estimates of the location of a given number of points in a given number of dimensions. The minimum number of estimates (or degrees of freedom) required is:

$$m(n - 1) - m/2(m - 1) \qquad (1)$$

where m is the number of dimensions and n is the number of points. Young postulated that with complete ordinal data, a ratio of comparisons elicited to this minimum necessary number of degrees of freedom can be used to provide an adequate indication of information recovery. According to Young, the size of the minimum necessary ratio is a function of the degree of error in the data. Young [19] found that, when he used normalized data $N(0,1)$ that was perturbed through the addition of a high degree of error $N(0,.5)$ added to each location before each distance calculation, adequate metric recovery with the use of MDS was obtained when this ratio was somewhere between 3.0 and 3.5.

In most cases, one of the parameters used in this measure, the dimensionality of a set of points, is not fully determinate or given. In our case, it has been shown that a two-dimensional solution of north-south and east-west coordinates adequately describes the data [2]. Thus, for our sample, the Young ratio is:

$$\frac{437}{95} = 4.60 \qquad \text{(from equation 1)}$$

According to this criterion, then, the number of comparisons used would seem more than adequate in determining a relatively "good" degree of metric determinacy. Spence and Domoney [16] pointed out that the Young ratio seems to be an adequate criterion for incomplete data designs.

A number of inadequacies involved in the use of this information in isolation have been presented, primarily by Hall and Young [4]. First, the shape or pattern of a given configuration of points affects metric determinacy. Second, it has not been shown conclusively that any measure such as stress—the common goodness-of-fit measure used in the Kruskal-Young-Shepard-Torgerson multidimensional scaling algorithm, KYST—is a good indicant of metric determinacy [9]. This is especially important since the chief concern of the location experiment was the recovery of metric approximations of cognitive data through the use of the paired comparison. Second, though MDS algorithms may be especially appropriate for the analysis of this type of data, the minimization of the stress optimization criterion associated with this method may not necessarily fit the aim of the analyst interested in the recovery of specific locational information [14].

Graef and Spence [3] added a third reservation when incomplete data is used. Utilizing a different indicant of goodness of fit, the correlation between simulated distances and distances between MDS and output locations, their simulation studies show that information recovery is enhanced to a greater extent by the inclusion of large distances than it is with the use of small distances. In this regard, Hall and Young's second objection should be kept in mind, since it will be shown later that this result may be related to the type of goodness-of-fit measure used by these authors. Last, when incomplete data are used, the Spence and Domoney results incicate that the Young ratio may not hold when distances are calculated among fewer than thirty points [16].

These arguments considerably expand the area required in the search for information recovery. They also indicate a need for either the direct testing of results against a configuration of expected locations or the theoretical derivation of probability density functions for results derived through multidimensional scaling under varying conditions of error. In terms of the first objection, it is fortunate that in this case there is a well-established model.

The Pattern of Data Collection in the Sampling Design

Given that a certain amount of data has been collected, what is the effect of the choice of sampling design on the ability to recover information about locational data? Three measures have been introduced in the literature that rely on aspects of sampling procedures significant to metric recovery. Two are reviewed by Spence and Domoney [16]. These measures tend to support the use of cyclic and other PBID designs. The design we chose for the location experiment is far from optimal under these criteria. It is most important, however, that we examine our design and the others suggested to ascertain how well they compare in terms of locational information recovery and the other measures reviewed by Spence and Domoney.

The first measure, "efficiency," was developed for paired comparison matrices by McKeon [10]. It is an extension of the efficiency criterion used in the analysis of variance. Specifically, it is the ratio of the average between case variance for a complete design and the average between case variance for an incomplete design. McKeon's efficiency is defined by:

$$\frac{d'(\text{com})s(\text{com})}{d'(\text{incom})s(\text{incom})} \qquad (2)$$

where $d'(\text{com})$ is the inverse of the mean number of comparisons of each object in a complete paired comparison design, $d'(\text{incom})$ is the inverse of the mean number of comparisons of each object in an incomplete design, $s(\text{com})$ is the population variance of a complete design, and $s(\text{incom})$ is the population variance of an incomplete design.

According to Spence and Domoney, this efficiency measure is highly related to aspects of connectedness in the incidence matrix of comparisons. The first of these mentioned by Spence and Domoney is "local" connectedness, which is related by them to the number of triangles or triplets of points directly interconnected in a sampling design. A triangle is formed when comparisons i,j; j,k; and i,k are all made, forming the graph:

The presence of a great number of triplets in an incomplete design is inversely related to "global" connectedness. The greater the "global" connectedness of a matrix, the greater the number of unique paths between any and all points i and j. Such relationships ensure greater "overall cohesiveness" among points in the sample. Spence and Domoney have calculated the number of triangles and McKeon's efficiency measure for a number of designs and have shown an almost invariant negative relationship between these two numbers [16].

The first column in Table 3 gives the McKeon efficiency measures for a sample of random designs, our design, the cyclic design and a design using only the largest distances. In this regard, the cyclic design tends to outperform, followed by the random designs, our design, and, last, the largest distances design. From an examination of Figures 2 through 5 (which show examples of the designs tested) the reasons are rather obvious. Our design (Figure 2) emphasizes a large degree of interconnectedness among a number of large cliques defined within the overlapping triangles of paired comparisons. Similarly, as Figure 5 shows, the large distance design leads to a great number of interconnections among a few points. (Compare points 23 and 30 to 26 and 32.) The number of paired comparisons involving each point tends to vary considerably, so that paths to some points are highly limited in number. Such a condition not only violates Kendall's first criterion, it also means that, though condition two probably holds, groups that are only minimally interconnected are easy to find [8].[1]

TABLE 3
Efficiency Measures Using Various Sampling Designs

Sampling Design	McKeon Efficiency	John Efficiency	Resistance
Our design	.799	.137	.408
Random (1)	.934	.117	.447
Random (2)	.929	.118	.444
Random (3)	.923	.119	.442
Largest distances	.742	.148	.355
Cyclic	.969	.112	.468

NOTE: Three different designs of randomly chosen distances were generated: random (1), random (2), and random (3). In subsequent tables only the first of these designs is used. Results using the other two designs are only marginally different and so are not reported.

A second measure highly related to McKeon's is proposed by John, Wolock, and David [6]. John's efficiency measure compares the variance of a given sample design to that of a randomized block design. This efficiency measure takes into account in a more obvious manner the possibility of differential utilization of objects within a paired comparison sampling design. Kendall's first criterion (that all sample points be used to the same extent in a sample design) can be rephrased in terms of this context, and its rationale easily is seen here. In this way, the unknown effect of weighting estimates differentially on the basis of the locations of certain well-used objects in a design is removed. Thus, in a cyclic design, all relative location estimates are based on an equal number of paired comparisons.

John's efficiency is the ratio of the mean variance estimates of the contrasts between points *i* and *j* using a given design to that of a random block design [6].

Last, Johnson and Van Dyk [7] used the denominator of the mean variance of a given sample design as an independent measure. They termed this an indicant of the "mean resistance" of a sample design. This concept is analogous to Spence and Domoney's idea of general connectedness. The lower the mean resistance, the greater the number of options and the ease of selecting options required in connecting two points in a sampling design. Van Dyk's measure performs in the same way for our sample designs as do the alternative indicants. (See column 3, Table 3.)

At this point, it is important to reintroduce a major determinant in the selection of our design. According to our past experiments, it is more important to emphasize certain objects and the relationships among them (largest and smallest distances, for instance). Given the division of our task into five subtasks, the effect of not including these extremes is the multiplication of each subset of paired comparisons estimates by a different constant (since extremum, upon which the comparisons were anchored, would vary over subsamples). Thus, one must weigh the advantages of a loss in "global" connectedness to a second factor introduced by the manner in which our data were collected.

Using these measures, it would seem that a number of advantages accrue in the use of a PBID design such as the cyclic. These advantages, though, must be weighed against the contingencies inherent in a given situation. The adequacy of a given design lies in how well it leads to the adequate representation of a given configuration of objects.

RECOVERY OF METRIC INFORMATION UNDER CONDITIONS OF ERROR

In order to examine how well differing sampling designs recover the type of information desired in our experiment, a simulation exercise was conceived. Distances based on our sample locations were perturbed on the basis of an error model introduced by Young [19]. Coordinates were first normalized ($N(0,1)$), removing the effects of differences in scale. Young's error perturbation involved adding a random, normally distributed error to each coordinate before the calculation of distances between sample points. The error perturbed distances, thus, have the form:

$$d(k,j) = (\Sigma(x(i,k) + e(i,k)) - (x(j,k) + e(j,k))^2)^{1/2} \qquad (3)$$

where e is the error perturbation distributed $N(0,s)$ and s is the error variance. Seven different error variances were used with this model, which is identical to that used by Young [19] and Issac and Poor [5].[2] (See Table 4.)

Because our task involved the use of an ordinal scale, the data were translated into a set of scale values via a simple transformation. The method of transformation involved defining equal interval distance classes and allocating sample distances after they were subject to error perturbation. Nine such classes were defined, graduated so that the members of the class of smallest distances received scale values of one and the members of the largest distance class received scale values of nine. The resulting sets of distances were first inserted into the nonmetric MDS algorithm KYST.

As is common practice in the scaling method, because of local minima problems, three starting configurations were used. These included: (1) the original sample locations before error perturbation, (2) coordinates calculated by using the error perturbed distances in the Torgerson optional starting configuration, and (3) the same scaling methodology as used for the second starting configuration, with the added complication that a four-dimensional solution was first generated. From this solution, principal components were determined and the component contributing the least variance was paired off. Again, a best nonmetric solution was calculated and the procedure was repeated until a two-dimensional solution was obtained.

Subsequently, squared correlations were calculated between the original locational coordinates and those derived using the error perturbed distances. Because an infinite number of rotated or reflected solutions equally satisfy the stress criterion inherent in an MDS solution, it was first necessary to rotate and reflect the scaling solutions to maximum congruence with the original coordinates. It was expected that, with the Young model, the solutions would be congruent to the point where the mean square error was close to the introduced error perturbation.

Table 4 lists the stress statistics using Kruskal's stress formula two for all four designs under the nine different error conditions and the two different distance types. Some interesting results emerge in Section I of Table 4. Within the Young model, under conditions of moderate and high error (variance of over .3), configurations based on the Torgerson starting configuration tended to yield better stress values. This occurrence was especially evident for the cyclic design.

From an examination of Section II of Table 4, it is evident that, when scaled distances were used, this tendency seemed to disappear for solutions based on distances perturbed with the Young model. A probable explanation emerges when the effect of our transformation is examined. In effect, these provide a looser set of distance constraints than the alternative normalized distance. Thus, the effects on stress of a number of the minor error perturbations were removed, since many distances grouped into a given class no longer led to violation of monotonicity constraints when error was added. (See the Shepard diagram, Figures 6 and 7.)

The squared correlations are given in Table 5. Despite rather similar stress statistics, the solutions based on the Torgerson starting procedures would seem rather different from those derived with the actual locations as the starting configuration. With our sample design, an error perturbed configuration could only be matched to the original locations when the actual locations were used as the starting configuration. Results using our design were almost too good—perhaps indicating that the distribution of constraints imposed with the framework of this design lift the possibility that solutions exist that are somewhat malleable to the shape of a given starting configuration. With the large distance design and the cyclic design, the solutions derived by using the Torgerson scaling procedure and subsequently reducing the dimensionality of the final configuration to produce a starting configuration resulted in rather aberrant results under conditions of moderate to high error, ($N(0,.2)$ to $N(0,.5)$). Finally, the random design tended to perform "best," yielding in almost all cases representative solutions no matter what the starting configuration.

TABLE 4
Stress Statistics for All Sampling Designs Using Normalized and "Scaled" Distances
and Various Sizes of Error Perturbation

Sampling Design	Error Size/Starting Configuration	Actual Distances	Torgerson Start/End = 2	Torgerson Start = 4/End = 2
		Section I — Normalized Distances		
Our design	.0	.000	.066	.005
	.1	.075	.102	.082
	.15	.104	.115	.123
	.2	.151	.158	.148
	.3	.191	.205	.213
	.35	.207	.217	.217
	.5	.269	.264	.295
Random (1) design	.0	.000	.004	.004
	.1	.076	.076	.076
	.15	.115	.115	.115
	.2	.128	.128	.128
	.3	.182	.181	.183
	.35	.203	.213	.213
	.5	.234	.242	.262
Largest distances design	.0	.000	.005	.005
	.1	.047	.047	.103
	.15	.067	.067	.100
	.2	.083	.083	.123
	.3	.088	.088	.098
	.35	.096	.096	.125
	.5	.098	.090	.105
Cyclic design	.0	.000	.004	.004
	.1	.081	.080	.080
	.15	.111	.111	.111
	.2	.158	.158	.185
	.3	.197	.179	.186
	.35	.209	.210	.189
	.5	.209	.192	.191

Results obtained by using distances perturbed with the Young model were subjected to hypothesis testing utilizing an analysis of variance format. Two sets of data were used: minimum stress statistics derived from KYST and the squared correlation statistics. The data was partitioned on the basis of: (1) the type of sampling design utilized, (2) whether or not the distances had been "scaled,"

TABLE 4 (Continued)

Sampling Design	Error Size/Starting Configuration	Actual Distances	Torgerson Start/End = 2	Torgerson Start = 4/End = 2
	Section II — Distances Transformed to Scale Values			
Our design	.0	.000	.049	.003
	.1	.025	.058	.033
	.15	.054	.063	.054
	.2	.094	.104	.099
	.3	.129	.145	.138
	.35	.148	.156	.150
	.5	.210	.211	.217
Random (1) design	.0	.000	.000	.003
	.1	.031	.031	.031
	.15	.059	.059	.059
	.2	.073	.073	.073
	.3	.129	.129	.129
	.35	.151	.153	.155
	.5	.181	.188	.184
Largest distances design	.0	.000	.005	.004
	.1	.025	.025	.025
	.15	.043	.044	.081
	.2	.063	.063	.114
	.3	.071	.071	.085
	.35	.085	.101	.125
	.5	.088	.088	.098
Cyclic design	.0	.000	.003	.004
	.1	.029	.029	.029
	.15	.054	.053	.052
	.2	.089	.089	.089
	.3	.136	.150	.162
	.35	.155	.157	.157
	.5	.187	.163	.172

and (3) the KYST starting configuration. Size of error, because of its monotonic property, was treated as a continuous independent variable. Since this variable was expected to affect stress size, it was designated as a covariate [11]. The interaction between the size of error and these other variables was very small. Note that there are 168 results based on 4 configurations x 2 scaled/nonscaled sets x 3 starting configurations x 7 levels of error. The null hypo-

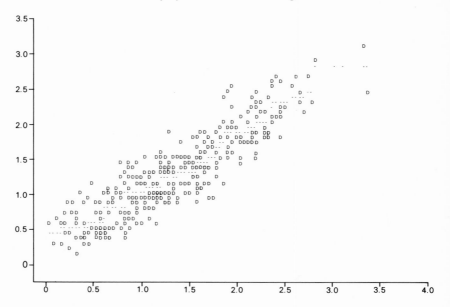

Figure 6. Shepard diagram for cyclic design; Young error model; "actual locations" start, error = .2; simulation results.

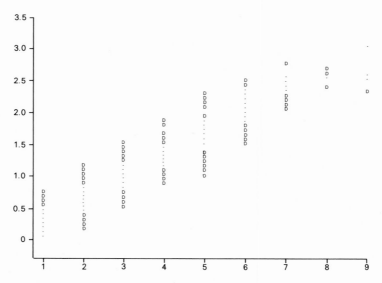

Figure 7. Shepard diagram for cyclic design; Young error model; "actual locations" start, error = .2; scaled distances; simulation results.

thesis took the form that there are no relationships between any of these factors nor their interdependencies and the size of stress or squared correlation statistics. A confidence level of 95% was used throughout this analysis.

Table 6 indicates the results for the stress statistics. The null hypothesis can be rejected with 95% confidence for the following: (1) the continuous independent variable, size of error; (2) the type of design used; (3) whether or not the distances used had been "scaled"; and (4) the combined effect of having the distances "scaled" and using different types of sampling designs. It should be noted that these effects "explain" roughly 83.3% of the adjusted sums of squares of the stress values.

In an examination of the means of the partitioned groups of stress statistics, it can be seen that the largest distance design did indeed produce the lowest stress statistics. Our design, in contrast, produced the highest.

"Scaled" values, as it should be expected, produced lower stress values than distances that were not so transformed. Again, examining Figures 6 and 7 makes it quite easy to conjecture about why this is so. It is obvious that the range of values falling within the boundaries defined by each distance class is much larger for the "scaled" values.

It is, however, difficult to explain the combined effect of design and the presence of "scaled" values. These effects are rather minor compared to the effect of increasing error (since only about 20% of the total explained variance is attributable to these sources).

Tables 8 and 9 list the more interesting results for the correlation statistics. The null hypothesis can be rejected with 95% confidence for the following: (1) the independent variable, size of error; (2) sample design; (3) type of starting configuration; and (4) the interaction between sample design and type of starting configuration. The last three relationships were not evident in the stress statistics.

Using the random design provides the highest mean squared correlation statistics; using the cyclic and largest distance designs, the lowest. These results are very different from those derived by using stress as the dependent variable. Lowest stress and highest squared correlation were not yielded by the same set of circumstances. This provides a good example of the poor role of stress by itself in determining the appropriateness of a given solution. Further, the use of "scaled" distances has little impact on the squared correlations.

TABLE 5
Correlation Statistics between Configurations Derived from Error Perturbed Distances
and the Original Locations

Sampling Design	Error Size/Starting Configuration	Actual Distances	Torgerson Start/End = 2	Torgerson Start = 4/End = 2
	Section I — Normalized Distances			
Our design	.0	1.00	.49	1.00
	.1	.99	.46	.91
	.15	.98	.42	.60
	.2	.97	.42	.93
	.3	.93	.53	.32
	.35	.89	.38	.35
	.5	.78	.36	.16
Random (1) design	.0	1.00	1.00	1.00
	.1	.99	.99	.99
	.15	.99	.99	.99
	.2	.96	.96	.96
	.3	.90	.88	.86
	.35	.87	.76	.76
	.5	.77	.56	.27
Largest distance	.0	1.00	1.00	1.00
	.1	.99	.99	.55
	.15	.97	.97	.42
	.2	.94	.94	.23
	.3	.88	.89	.41
	.35	.87	.87	.32
	.5	.81	.67	.45
Cyclic design	.0	1.00	1.00	1.00
	.1	.99	.99	.99
	.15	.98	.98	.98
	.2	.96	.96	.16
	.3	.90	.39	.09
	.35	.81	.65	.04
	.5	.44	.12	.04

Last, the type of starting configuration definitely affects the closeness of the results to the actual locations, since a starting configuration yielded results closer to the actual locations. Given that lowest stress was not significantly affected by the starting configuration used, this leaves one rather suspicious. The incomplete data design

TABLE 5 (Continued)

Sampling Design	Error Size/Starting Configuration	Actual Distances	Torgerson Start/End = 2	Torgerson Start = 4/End = 2
Section II — Distances Transformed to Scale Values				
Our design	.0	1.00	.62	.99
	.1	.99	.61	.93
	.15	.98	.42	.97
	.2	.97	.41	.32
	.3	.92	.57	.37
	.35	.89	.40	.33
	.5	.80	.37	.22
Random (1) design	.0	1.00	.99	.99
	.1	.99	.99	.99
	.15	.99	.98	.98
	.2	.94	.94	.94
	.3	.92	.89	.90
	.35	.89	.76	.73
	.5	.82	.69	.61
Largest distance	.0	1.00	.97	.98
	.1	.99	.99	.99
	.15	.98	.98	.62
	.2	.96	.96	.24
	.3	.91	.91	.45
	.35	.92	.62	.32
	.5	.85	.59	.30
Cyclic design	.0	1.00	.99	.99
	.1	.99	.98	.99
	.15	.97	.95	.96
	.2	.96	.96	.96
	.3	.91	.38	.40
	.35	.89	.70	.11
	.5	.70	.20	.02

utilized seemed to provide an insufficient number of constraints on what the MDS solution could be. Thus, similar low stress statistics for a given set of paired comparisons existed, while the solutions associated with these statistics would seem to be very different. The interaction between design and type of starting configuration would seem to suggest that this problem can be partially remedied by the use of an appropriate sampling design.

TABLE 6
Analysis of Variance Results for Stress Statistics: Sources of Variation

Source	Sum of Squares	Degrees of Freedom	Mean Square	F	Significance of F
Covariate:					
Size of error	.558	1	.558	677.31	.001
Main Effects:					
Type of design	.145	3	.027	32.84	.001
Normalized/scaled	.062	1	.062	75.71	.001
Starting configuration	.002	2	.001	1.24	.293
Two-way Effects:					
Design x normalized/scaled	.008	3	.003	3.37	.020
Design x starting configuration	.005	6	.001	1.10	.364
Normalized configuration/scaled x starting configuration	.000	2	.000	0.06	.999
Three-way Effects:					
Design x normalized/scaled x starting configuration	.000	6	.000	0.03	.999
Explained	.717	24	.030	36.24	.001
Residual	.118	143	.001		
Total	.835	167	.005		

Approximately 66.5% of the variance of the squared correlation statistics was "explained" by these variables. This was not as high as the percentage "explained" by using stress as the dependent variable. Partitioning variance on the basis of design, configuration, and the interaction of these two accounted for 34% of the variation of the squared correlations—a major portion of variation. It would seem, then, that the importance of such considerations becomes magnified when this indicant of metric determinacy is examined.

A major problem is also evident here. Contrasting and comparing the results of Tables 6 through 9 make it evident that lowest stress is not necessarily a guarantee of greater information recovery using this squared correlation measure, especially under conditions of incomplete data and high error. Examining, for example, the cyclic design solutions in which the error variance used is greater than .3 makes this point quite obvious. It is evident that lowest stress and greatest information recovery do not go hand in hand.

TABLE 7
Analysis of Variance Results for Stress Statistics:
Deviations from Mean

Variable	N	Mean Deviation
Design:		
Ours	42	.02
Random (1)	42	.01
Large distances	42	−.04
Cyclic	42	.01
Normalized/Scaled:		
Normalized	84	.02
Scaled	84	−.02
Starting Configuration:		
Actual distances	56	.00
Torgerson start and end = 2	56	.00
Torgerson start = 4 and end = 2	56	.00
Grand mean: .11		

Figures 8 through 11 are examples of lowest stress solutions achieved by using specific designs and error levels. Figures 8 and 9 show, in addition, the displacements between the model locations of Figure 1 and these solutions.

Figure 8 shows the lowest stress solution, given "our" design and an error level of .5. This should be compared with Figure 9, which shows the lowest stress solution achieved with a cyclic design and a lower error level of .3. (See Table 4 for stress statistics.) Both yield squared correlation coefficients of roughly .35 when compared to the model. (See Table 5.) Figure 8 shows a number of moderately large displacements. In contrast, Figure 9 shows points 7, 22, 30, and 43 dramatically shifted across the mapping from their original locations.

Figures 10 and 11 indicate that the largest distance design solutions seemed to display an inordinate amount of clustering, leading to a number of seemingly degenerate solutions. It is obvious that relationships among places close to each other are not well represented. Since this does not affect the square correlation coefficient as much as a departure with the same relative impact where places are far apart, a problem of priorities is evident. Is the

TABLE 8
Analysis of Variance Results for Squared Correlation Statistics: Sources of Variation

Source	Sum of Squares	Degrees of Freedom	Mean Square	F	Significance of F
Covariate:					
Size of error	4.227	1	4.227	159.02	.001
Main Effects:					
Type of design	1.148	3	0.383	14.40	.001
Normalized/scaled	0.039	3	0.039	1.47	.290
Starting configuration	2.431	2	1.216	45.74	.001
Two-way Effects:					
Design x normalized/scaled	0.027	3	0.009	0.33	.999
Design x starting configuration	1.507	6	0.251	9.45	.001
Normalized/scaled x configuration	0.026	2	0.013	0.49	.999
Three-way Effects:					
Design x normalized/scaled x starting configuration	0.052	6	0.009	0.33	.999
Explained	9.457	24	0.394	14.83	.0
Residual	3.801	143	0.027		
Total	13.258	167	0.079		

inordinate clustering provided by the largest distance design a real advantage in any other facet of analysis than the tendency to conserve high correlation coefficients?

With the scaled distances, results were almost identical. One interesting result was the slightly better degree of information recovery (as indicated by the squared correlation statistics) obtained under conditions of high error with the Young model.

CONCLUDING STATEMENTS

It is obvious that more thorough experimentation needs to be carried out using various incomplete designs and levels of data exclusion under conditions of error. Despite the apparent advantage of the cyclic design and our own design, it is not clear from the results reported here that these advantages are substitutes or major compensations for loss of information.

Especially worrisome are the stresss results and their divergent

TABLE 9
Analysis of Variance Results for Squared Correlation
Statistics: Deviations from Mean

Variable	N	Mean Deviation
Design:		
Ours	42	−.10
Random (1)	42	.13
Large distances	42	.01
Cyclic	42	−.04
Normalized/Scaled:		
Normalized	84	−.02
Scaled	84	.02
Starting Configuration:		
Actual distances	56	.16
Torgerson start and end = 2	56	−.02
Torgerson start = 4 and end = 2	56	−.13

Grand mean: .77

trends when compared to the squared correlation statistics given this set of test locations. When data are incomplete and when error is present, the possibility exists that the data are malleable into differently shaped configurations that yield similar or identical stress statistics. As the example of the large distances sample design shows, this could be a side effect of the data design itself.

Thus, without prior information concerning what a configuration should look like, it is rather dangerous to interpret lowest stress results based upon the amount of information used in this research. Even Shepard's comments concerning the need to emphasize the relative structure of an MDS configuration is questionable given our experiment, which used this amount of missing information because of such tendencies as the reflection of points [14]. (See again Figure 9, especially.)

In this research, the presence of a model brings forth some assurance but also leaves a somewhat uneasy feeling. In examining individuals' cognitive configurations, an hypothesis remains that these bear a resemblance to the "actual" map locations. Using a set of malleable distance estimates that can be fit with the same stress to more than one pattern leaves the acceptance of this hypothesis in doubt, since any configuration then becomes one of a

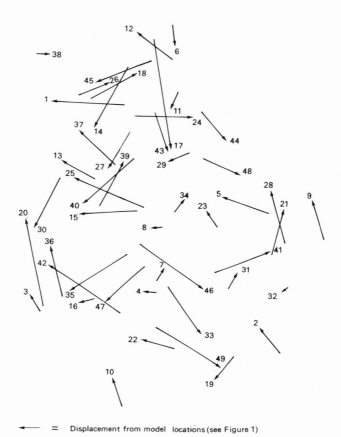

— = Displacement from model locations (see Figure 1)

Figure 8. Lowest stress solution using a cyclic sampling design and an error level of .5.

number of results. Under conditions in which the type of evaluation is unknown and the ways in which relative distances are far from completely understood, a great deal of care should be utilized in interpreting this type of MDS result.

NOTES

1. It would seem advantageous in this regard to actually compute the number of unique paths between points requiring different numbers of intermediate links. Each of these, using the terminology of graph theory, defines a distance. It would seem that for matrices exhibiting a high degree of "global" connectedness, the number of paths of

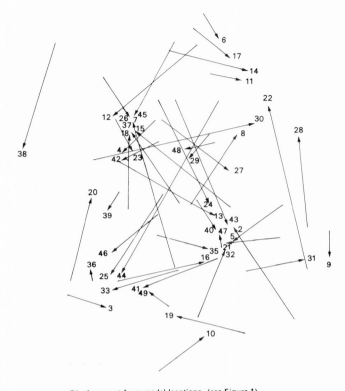

——— = Displacement from model locations (see Figure 1)

Figure 9. Lowest stress solution using a cyclic sampling design and an error level of .3.

varying distance would not differ as much as for matrices that have a great number of triangles. Further, it would be expected that, when the number of comparisons and the number of stimuli used are equal, greater numbers of large distance interconnections would exist. Using Parthasarathy's algorithm for determining the number of unique paths of varying distance, this hypothesis could be tested [12].

2. An alternative, multiplicative model introduced by Wagenaar and Padmos was also used [18]. For a discussion of these results, see Spector's dissertation [15].

REFERENCES

1. David, H. A., 1963. *The Method of Paired Comparisons*. London: Griffin.
2. Golledge, R. G., V. L. Rivizzigno, and A. N. Spector, 1976. "Analytic Methods for Determining and Representing Cognitive Configurations of a City." In R. G. Golledge

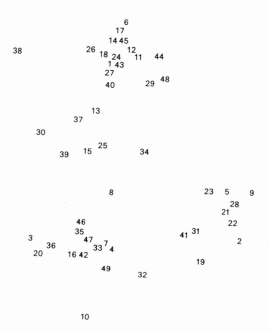

Figure 10. Lowest stress solution using a largest distance design and an error level of .2.

and J. N. Rayner (eds.), *Cognitive Configurations of the City* (vol. 2). Columbus, Ohio: Ohio State University Research Foundation.

3. Graef, J., and I. Spence, 1976. "Using Prior Distance Information in Multidimensional Scaling," a paper presented at the Annual Meetings of the Psychometric Society, April 1-3, 1976, Murray Hill, N.J.

4. Hall, R. G., and F. W. Young, 1975. *Nonmetric Multidimensional Scaling: A Volumetric Index of Partial Metric Determinacy.* Chapel Hill, N.C.: University of North Carolina, L. L. Thurstone Psychometric Laboratory Research Report #138.

5. Isaac, P. D., and D. D. S. Poor, 1974. "On the Determination of Appropriate Dimensionality in Data with Error," *Psychometrika*, 39, 91-110.

6. John, J. A., F. W. Wolock, and H. A. David, 1972. *Cyclic Designs.* Washington, D.C.: National Bureau of Standards Applied Mathematical Series #62.

7. Johnson, R. M., and G. T. Van Dyk, 1974. "A Resistance Analogy for Pairwise Experimental Design." Unpublished manuscript, Market Facts, Chicago.

8. Kendall, M. G., 1955. "Further Contributions to the Theory of Paired Comparisons," *Biometrics*, 11, 43-62.

9. Kruskal, J., F. Young, and J. Seery, no date. "How to Use KYST, a Very Flexible Program to Do Multidimensional Scaling." Unpublished report, Bell Laboratories, Murray Hill, N.J.

10. McKeon, J. J., 1960. *Some Cyclical Incomplete Paried Comparison Designs.* Chapel Hill, N.C.: University of North Carolina, L. L. Thurstone Psychometric Laboratory Research Report #24.

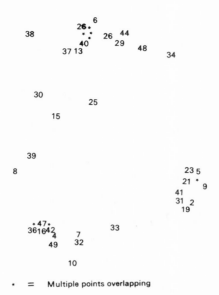

* = Multiple points overlapping

Figure 11. Lowest stress solution using a largest distance design and an error level of .5.

11. Nie, N. H., C. H. Hull, J. G. Jenkins, K. Steinbrenner, and D. H. Brent, 1975. *Statistical Package for the Social Sciences* (2nd ed.). New York: McGraw-Hill.
12. Parthasarathy, K. R., 1964. "Enumeration of Paths in Digraphs," *Psychometrika*, 29, 153-65.
13. Rivizzigno, V. L., 1976. "Cognitive Representations of an Urban Area." Ph.D. dissertation, Department of Geography, Ohio State University, Columbus, Ohio.
14. Shepard, R. N., 1974. "Representation of Structure in Similarity Data: Problems and Prospects." *Psychometrika*, 39, 373-421.
15. Spector, A. N., 1978. "An Analysis of Urban Spatial Imagery" Ph.D. dissertation, Department of Geography, Ohio State University, Columbus, Ohio.
16. Spence, I., and D. W. Domoney, 1974. "Single Subject Incomplete Designs for Nonmetric Multidimensional Scaling," *Psychometrika*, 39, 469-90.
17. Torgerson, W. S., 1958. *Theory and Methods of Scaling.* New York: John Wiley and Sons.
18. Wagenaar, W. A., and P. Padmos, 1971. "Quantitative Evaluation of Stress in Kruskal's Multidimensional Scaling Technique," *British Journal of Mathematical and Statistical Psychology*, 24, 101-10.
19. Young, F. W., 1970. "Nonmetric Multidimensional Scaling: Recovery of Metric Information," *Psychometrika*, 35, 455-73.

Chapter 1.5 Considerations in the Selection of Stimulus Pairs for Data Collection in Multidimensional Scaling

Paul D. Isaac

A recurrent problem in data collection for multidimensional scaling is the rapid increase in the number of distances $\binom{n}{2}$ as the number of stimuli (n) is increased. For designs involving direct judgment of similarity, the solution seems to be to reduce the number of pairs actually judged. Two approaches previously taken have been the use of interactive scaling [10], which involves on-line participation of a subject with rational selection of pairs based on previous responses, and the use of cyclic designs to delete pairs a priori [6]. Neither of these approaches is altogether satisfactory — interactive scaling because of the on-line requirement and cyclic designs for reasons discussed below.

If the number of stimulus pairs actually judged is to be reduced, it seems that there are at least two issues that must be considered. One issue concerns the distances that are most important to an MDS algorithm in determining a configuration. The second is the issue of the reliability of the judgments of particular distances (e.g., is the information on short or long distances the most trustworthy?). (I am using the term *judgments*, but, more generally, any measure of pairwise similarity could be involved — confusions, reaction times, correlations, etc.) The rest of this paper will deal with these two issues.

DISTANCES IMPORTANT FOR RECOVERING
A CONFIGURATION

As an initial conjecture, I hypothesized that not all distances are equally important in determining a configuration via nonmetric MDS. In particular, it would seem that the short and long distances would be most important and medium distances less so. The long distances should serve to outline the configuration, and the short ones, to provide local detail. Interestingly, a study by Graef and Spence [2] provides partial support for this conjecture. (The details of this paper appeared after an initial draft of my paper, but I will refer to it from time to time.) In particular, they constructed configurations by sampling thirty-one points within a unit circle and adding error (monotone with distance) to the distances between points to provide simulated dissimilarity measures. In each case, they deleted one-third on the data values—the small, medium, or large third—prior to obtaining an MDS solution using TORSCA-9. They found that deletion of the large distances had the most dramatic deleterious effect on recovery of the true configuration, that is, that the large distances were most important for recovery of the true configuration. No compelling advantage of small versus medium distances was apparent. Thus, it seems that any data-collection scheme that would eliminate some pairs from consideration should not eliminate those involving large distances. It is not obvious, though, that my conjecture concerning small distances is correct or relevant (particularly when considering, in addition, the scaling folklore on the untrustworthiness of local information in scaling solutions).

RELIABILITY OF JUDGMENTS

It seems obvious that not all distances will be judged (or responded to in some way) with equal accuracy. Indeed, Monte Carlo studies with complete [3, 5, 9] and incomplete data [2, 6] have reflected this in error functions that have made error increase with increasing true distance. However, it also seems that this type of monotone increasing error function will not be appropriate as a simulation of all tasks relevant to MDS. A number of studies bear this out empirically.

Golledge and his co-workers [1] have made extensive use of

direct judgments of distance between locations in cities. They find that the judgments of long distances tend to be most in error. There certainly are a number of reasons why this might be the case.

Young [8] made use of mean reaction times (RTs) as inverse measures of distance. There tend to be more errors made in responding to small differences than to large. The net effect on mean RT is to make the standard error larger for these small differences because there are fewer usable correct RTs. Thus, an appropriate error function would be inversely monotonic with true distance. (It should be noted that fast reaction times are more often in error in semantic research, also. The possible effects on studies using MDS in this area (e.g., [4]) has not been explored.)

In a previously unreported study, I had twelve graduate students make dissimilarity judgments of all pairs of ten political candidates on a twenty-point scale. After one week, they repeated these judgments. This gave two sets of twelve ten-by-ten matrices, which were input into KYST. Two-dimensional solutions were obtained in a number of ways: common monotonic regression for all subjects on each replication separately; separate regressions for each subject on each replication (split by groups); solution based on means of judgments in each replication; and solution based on mean of judgments for both replications together. In each case, the solutions were used to determine small, medium, and large thirds of distances. The judgments corresponding to these distances were then correlated from replication 1 to replication 2. The general finding was that the replication 1-replication 2 correlation (Pearson or Spearman) tended to be lowest for the judgments of medium distances (as given in the particular two-dimensional solution). The data for this study is reported in Table 1. Thus, in this case it seems that error is not monotone with distance—a conjecture previously made by Wish [7] with respect to category scale responses. This might result from nonsymmetric error functions on the ends of the category scale and symmetric ones in the center. (Differences in range or variance for the small, medium, and large judgments would not account for the obtained results, although details would differ somewhat.) These empirical findings suggest that a monotone increasing error function does not accurately simulate all kinds of judgments.

Another consideration that seems not to have been explored is the effect of the type of configuration generating the judgments.

TABLE 1
Correlations of Judgments Made One Week Apart: Correlations Computed for Small, Medium, and Large Distances

Source of Distances	Small				Medium				Large			
	F	N_r	$\bar{\rho}$	N_ρ	F	N_r	$\bar{\rho}$	N_ρ	F	N_r	$\bar{\rho}$	N_ρ
Common regression												
Replication 1	.715	3	.673	3	.622	7	.606	6	.792	0	.767	1
Replication 2	.727	3	.708	2	.671	6	.600	6	.772	1	.735	2
"Split-by-groups" solution												
Replication 1	.716	2	.673	3	.543	7	.616	6	.788	1	.754	1
Replication 2	.702	1	.651	3	.628	8	.568	6	.741	1	.694	1
Solutions from mean of judgments												
Replication 1	.651	3	.603	3	.501	6	.502	6	.762	1	.724	1
Replication 2	.516	6	.510	5	.657	3	.615	4	.656	1	.655	1
Replication 1 + replication 2	.649	2	.581	4	.462	7	.448	6	.696	2	.693	0

NOTE: Correlations of all distances: $\bar{r} = .808$, $\bar{\rho} = .783$.

F = Mean (of ten subjects) Pearson product moment correlation of replication 1 judgment with replication 2 judgment.

$\bar{\rho}$ = Mean Spearman *rho*.

N_r = Number of subjects for whom the Pearson r in that distance category (small, medium, large) was the smallest of the three for that subject.

N_ρ = Number of subjects for whom the Spearman ρ in that category was the smallest of the three for that subject. The N_r and N_ρ sum to ten across small, medium, and large distance categories.

The impact of the type of error function is likely to depend on the distribution of true distances (which will depend on the type of configuration). If there are few long distances (i.e., a positively skewed distribution of distances), then the error function may be irrelevant; if there are many, then it may be quite important. I know of no data directly relevant to this point.

IMPLICATIONS FOR SELECTION OF PAIRS TO BE JUDGED

I have suggested that two factors be considered when eliminating judgments from consideration in collecting data: the importance of the distance being judged and the reliability of the distance information. Graef and Spence [2] give support to the notion that (information on) long distances are necessary to obtain a satisfactory solution. Further, their error function was such as to make the long distances have the most error. (It might be interesting to examine the effect of type of configuration and distance distribution on their findings, especially relative to the importance of medium and small distances.) Thus, it seems that any procedure for data collection should be sure to include them. It is also clear that including only long distances for judgments will not ensure that all stimuli are properly located. A choice of additional distances to include would seem to depend on the task involved and the distribution of distances.

There are practical problems with these suggestions. The large distances are obviously important, but they must be determined before judgments can be made. This is a relatively minor problem and intuition would suggest a number of possible solutions. More important, however, is that judgment of only large distances may leave some stimuli that are never judged at all. Thus, some procedure that would ensure that all stimuli are judged is required.

PROCEDURES FOR SELECTION OF DISTANCES TO BE JUDGED

Spence and his colleagues have examined at least two kinds of procedures for reducing the number of judgments: cyclic designs and procedures that identify large distances. I would call these one-stage and two-stage procedures, respectively. A one-stage procedure determines which pairs to include prior to any data collection. A two-stage procedure involves preliminary data collection to determine which pairs to include in the second stage. I suspect that there may be some advantages to the two-stage procedures in that they can be designed to take advantage of what is known about which distances are important and which are reliably judged. The one-stage procedure cannot do this. Graef and Spence [2] have suggested several two-stage procedures that would ensure inclusion of large distances. Another procedure that might be useful is the following. *Stage 1:* Present each stimulus, in turn, to the subject as a target. Ask subject to select the k stimuli most like the target and the k least like it; call it "pick k/k," where $k = 2$, 3, ... Construct a design matrix from this initial stage by putting an X in the $2k$ entries thus chosen in each row of the n x n square matrix of a possible stimulus pair. *Stage 2:* Collect pair comparisons on only those pairs indicated by an X in the cell of the design matrix, eliminating duplications where they occur (i.e., collect only i,j, where both i,j and j,i are indicated). This procedure would be called for if it were suspected that medium distances have most error. Although admittedly it does not omit all medium distances, it does ensure that small and large distances will be included and further that all stimuli will be judged at least $2k$ times. Another characteristic of this procedure is that it is likely to be sensitive to the shape of the configuration in that the number of pairs finally included in the design will be a function of the type of configuration that should result. This could not occur in any one-stage procedure.

A BRIEF SIMULATION STUDY

In order to examine the characteristics of the "pick k/k" procedure, a small computer simulation was run. Three different types of two-dimensional configurations (forty points in each) were constructed: solid diamond, outline diamond, and outline circle. These were chosen to give both dense and shell-type configurations and two degrees of elongation (i.e., some and none), as well as differing distributions of distances. (See Figure 1.) From these basic types, different size configurations were selected: twenty-four, thirty, and forty points for the solid diamond and circle and twenty-four and thirty for the outline diamond. The parameter k (in "pick k/k") was varied from two to seven. This study involved no error function since the interest was primarily in the value of k needed to reconstruct the configuration adequately. The design of this study was not a completely crossed factorial, due to its exploratory

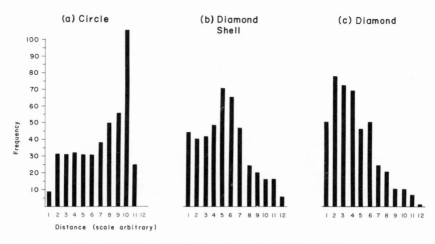

Figure 1. Thirty-point configurations: distributions of distances.

nature and due to the fact that the results obtained could reason-ably be extrapolated in certain obvious cases.

Incomplete data matrices were constructed for each size-type configuration by using the "pick k/k" (pick largest and smallest k in a given row) procedure on the true distances. In addition, the cyclic design, type II, of Spence and Domoney was used to select one-third of the distances (two-third deletion) for purposes of comparison. Degrees of freedom ratios (as discussed by Spence and Domoney [6]) were in some cases marginal with this proce-dure. However, it was felt that, if in fact long and short distances were most important in reconstructing a configuration, previous lower limits on the degrees of freedom ratio might be conservative-ly high since they did not take into consideration the differential importance of distances.

Incomplete matrices thus constructed were input into KYST, with a two-dimensional TORSCA start in each case. (Some ran-dom starts were tried, but the solutions were usually inferior to those using TORSCA, and consequently they were not used for all incomplete matrices.) Table 2 presents the results of these analyses: stress values, correlation of recovered and true distances, and number of data values actually selected by the "pick k/k" and cyclic designs. Figure 1 presents the distributions of distances in-volved in each configuration and for the different incomplete so-lutions.

TABLE 2
Stress Values and Correlations of Recovered and True Distances

Number of Points	k for "Pick k/k"	Circle			Diamond			Diamond Shell		
		Stress	r*	Number of Data Values	Stress	r	Number of Data Values	Stress	r	Number of Data Values
24	2	.009	.959	66	.010	.955	75	.000	.833	69
	3	.006	.991	88	.008	.990	104	.010	.973	107
	4	.009	.998	107	.009	.990	136	.005	.983	133
	5	—	—	—	.005	.997	163	—	—	—
	Cyclic 1/3	.008	.055	96	.010	-.036	96	.021	.383	96
30	2	.008	.931	75	.009	.909	95	.006	.858	91
	3	.009	.989	106	.010	.923	137	.010	.884	137
	4	.007	.997	134	.015	.983	176	.009	.983	177
	5	.006	.999	160	.005	.929	207	.004	.998	211
	Cyclic 1/3	.037	.083	155	.050	.874	155*	.028	.963	155
40	2	.007	.913	92	.010	.830	125	.008	.805	119
	3	.010	.985	147	.010	.830	182	.010	.856	181
	4	.009	.997	180	.010	.983	234	.009	.955	238
	5	.004	.998	221	.007	.995	185	.010	.990	291
	6	.003	.999	266	.005	.996	338	.009	.998	341
	7	—	—	—	.004	.999	386	.010	.997	387
	Cyclic 1/3	.004	.999	260	.009	.994	260	.048	.972	260

*r = Pearson correlation between true and reconstituted distances.

The results of this simulation are quite predictable: the more data values included, the better solution—as k increases, so does the goodness of the recovery. The one qualification to this involves the cyclic designs. Four of the nine cyclic solutions were unacceptable when evaluated by r (Pearson r of true and reconstructed distances). The fifth was marginally acceptable on the basis of r (.874) but actually unacceptable based on visual inspection (it did not look like a diamond). In the remaining cases, the cyclic solutions fit about as well as the "pick k/k" solutions with comparable numbers of data values. It should be noted that all of these solutions involved perfect data (i.e., no error).

Stress values for all solutions were no greater than .05, but, with three exceptions, the stress values were highest for the cyclic solutions to a given configuration. It is quite clear that, with incomplete data, stress is not a very trustworthy index of goodness of fit (e.g., for the twenty-four-point diamond, stress is .01, but r equals $-.036$).

One hypothesis concerning the "pick k/k" procedure was partially confirmed; the number of pairs actually picked depended on the type of configuration. There were fewer pairs (data values) selected for the circle than for either of the diamonds. However, the diamond and the diamond shell were not systematically different in the number of pairs picked.

CONCLUSIONS

These results support those previously presented by Spence [5] — in fact, long distances are useful in reconstructing configurations. Various procedures are possible for selecting long distances prior to final data collection. One—the "pick k/k" procedure—is presented here. Obviously, this could be modified to "pick k" (the k stimuli most different from a given stimulus) to ensure that all stimuli are judged (thus circumventing the problem mentioned above with respect to long distances).

The impact of different distance functions on solutions remains to be examined in detail. Graef and Spence [2] have results that seem to suggest that it is not worth looking at other error functions—long distances seemed to be best in the worst possible situation, i.e., with error monotonically increasing with distance. It would be of interest, however, to know the functional effect of the error, that is, how many reversals of order in the long distances actually occurred as a result of this error. The functional effect of error is likely to be dependent on the distribution of distances. Thus, varying configurations to give different distributions of distances while simultaneously varying the error function would be an appropriate way to address the issue of the impact of reliability on reduction of observations for MDS.

Thus, data on long distances seem necessary for adequate MDS solutions. The dependence of which and how many pairs to be selected on task (implying different error functions) and expected configuration remain to be explored.

REFERENCES

1. Golledge, R., J. Rayner, and V. Rivizzigno, 1974. "The Recovery of Cognitive Information about a City," a paper read at North American Classification Society and Psychometric Society Meetings, Iowa City, Iowa, 1974.
2. Graef, J., and I. Spence, 1976. "Using Prior Distance Information in Multidimensional Scaling," a paper presented at the Annual Meetings of the Psychometric Society, April 1-3, 1976, Murray Hill, N.J.
3. Isaac, P. D., and D. D. S. Poor, 1974. "On the Determination of Appropriate Dimensionality in Data with Error," *Psychometrika*, 39, 91-109.
4. Rips, L., E. Shoben, and E. Smith, 1973. "Semantic Distance and the Verification of Semantic Relations," *Journal of Verbal Learning and Verbal Behavior*, 12, 1-20.
5. Spence, I., 1972. "A Monte Carlo Evaluation of Three Nonmetric Multidimensional Scaling Algorithms," *Psychometrika*, 37, 461-86.
6. Spence, I., and D. Domoney, 1974. "Single Subject Incomplete Designs for Nonmetric Multidimensional Scaling," *Psychometrika*, 39, 469-90.
7. Wish, M., 1972. "Notes on the Variety, Appropriateness, and Choice of Proximity Measures," a paper for the Workshop on MDS, June 7-10, 1972, Bell Laboratories, Murray Hill, N.J.
8. Young, F. W., 1970. "Nonmetric Scaling of Line Lengths Using Latencies, Similarity, and Same-Different Judgments," *Perception and Psychophysics*, 8, 363-69.
9. Young, F. W., 1970. "Nonmetric Multidimensional Scaling: Recovery of Metric Information," *Psychometrika*, 35, 455-73.
10. Young, F. W., and N. Cliff, 1972. "Interactive Scaling with Individual Subjects," *Psychometrika*, 37, 385-415.

Chapter 1.6 The Interface between the Types of Regression and Methods of Collecting Proximity Data

Phipps Arabie and *Sigfrid D. Soli*

The advent of nonmetric multidimensional scaling as it appeared to researchers in the area [31,32,36], consisted of two major leaps forward: (1) the input data were only viewed as being measured on an ordinal scale when Kruskal's [17] monotone regression was substituted for earlier methods that made heavier demands of the data, and (2) the cumbersome method of complete triads [38] was no longer the only truly respectable procedure for collecting the proximities matrix. In fact, any consistent method that yielded a partial ordering of all the interstimulus proximities, or a major subset thereof, became admissible.

Thus, the investigator employing nonmetric multidimensional scaling ostensibly received carte blanche to use any of a variety of methods for collecting data, followed by a statistical technique that only assumed monotonicity. Among the tasks suitable for obtaining proximity matrices are category judgments of pairwise similarities, ratio comparisons of paired differences [24], "direct"

For their helpful remarks during discussions of this work, we are indebted to S. A. Boorman, J. J. Jenkins, J. B. Kruskal, P. R. Levitt, R. H. Shepard, C. E. Speaks, and W. Strange. We also benefited greatly from the remarks of T. R. Edman and G. P. Widin on an earlier version of this paper. This research was supported by National Science Foundation grants SOC 76 24512, SOC 76 24394, and GB-35703X, as well as National Institute of Child Health and Human Development grant HG-01136. The last two grants were made to the Center for Research in Human Learning at the University of Minnesota.

ratio scaling [8, 13], sorting [20], balanced incomplete subsets of triads [2, 19], and other methods. The considerations that the experimenter faces in choosing among these methods are well known: sophistication required of subjects (e.g., infants versus adults), number of trials involved, amount of time required, inherent difficulty of the task, et cetera.

But, in addition to these familiar considerations, there is the fact that some methods generally score a better payoff than others with respect to producing useful and interpretable nonmetric scaling solutions. For instance, there have been few successful interpretations of nonmetric solutions for two-way confusion matrices, aside from Shepard's [33, 34, 35] analyses of the Miller and Nicely [21] data for confusions between consonant phonemes and other data for confusions between Morse Code signals [15, 23, 26, 27]. Some authors (e.g., [10]) have even disparaged the overall utility of confusion matrices.

There are several possible explanations for the difference in success rates between methods: for example, sorting [20] yields overwhelmingly tie-bound data, and confusion matrices are ofter rather sparse. Most of these problems are apparent not only from in-regression function relating recovered metric distances to the input proximities data. For different data sets all collected by the same general experimental procedure, it is reasonable to expect that the scatter plot for each set of data should reflect in common (1) the drawbacks inherent in the experimental procedure (e.g., too many ties) and (2) any hallmarks of the judgmental process involved. (The first item probably will strike the reader as somewhat obvious; the second is presently being asserted as one of the foci of this paper.)

To the extent that the regression function for a given experimental procedure is consistently determined by these two considerations, one can seek to identify a prototypical regression function for the scatter plot of recovered distances versus inputs proximities. It should be noted that the scatter plot evidence for prototypical regression forms is one of the benefits of nonmetric multidimensional scaling. The classical techniques [38] generally assumed without any prior tests that the regression function was linear and proceeded from there. A more appropriate strategy would be to identify (via nonmetric methods) the prototypical form of regression specific to the experimental procedure. The data analyst can then perform the appropriate transformation so that the regression function is rectified, and linear regression, with its advantages over monotone regression, can justifiably be used.

The plan of this paper is, for illustrative purposes, to use the approach just outlined for a set of so-called "confusion data." (Psychologists gather confusion matrices from errors made by subjects performing various recognition and learning tasks.) We will compare the results of our approach with those of other investigators who have used different methods on the same set of data. However, it should be emphasized that the approach advocated here is not restricted to confusion matrices. As another case, consider correlation matrices. In general, such matrices have not yielded much in the way of new information when used as input for nonmetric scaling. Even some of the most competent studies [e.g., 9] have failed to yield results that had not been previously demonstrated by (metric) factor analysis. Although computing correlations between columns or rows of a rectangular matrix is clearly different from the common judgmental process assumed to underlie confusions, it still seems plausible to argue that the nature of product-moment correlation causes such matrices to be in some sense homologous. The classic paper by Cronbach and Gleser [7] suggests that otherwise heterogeneous matrices become similarly cloaked when correlations are computed upon the rows or columns. The present argument implies that it is worthwhile to seek a prototypical regression function for correlation matrices, with one natural candidate being Fisher's r-to-Z transformation (i.e., arctanh r).

EXPONENTIAL DECAY FUNCTIONS
FOR BEHAVIORAL PROCESSES

Background

The unsophisticated user of nonmetric multidimensional scaling often has no strong interest in the scatter plot of the regression function. Indeed, that part of the output is often overlooked entirely, in spite of the useful hints that it can give concerning local minima, degenerate solutions, and rational functions for regression. However, that function has long been emphasized in the animal learning tradition as the stimulus generalization gradient [11, 22, 25] depicting the unidimensional psychological distance between pairs of stimuli that are occasionally confused. (Interest in the function from animal studies preceded by several decades the advent of scaling programs for reconstructing satisfactory maps of the stimulus space.) The long-standing research activity in stimulus generalization accounts for the fact that of the facets of judgment considered earlier for collecting proximities, only for confusability is there a well-developed theory [29, 30] to which the prototypical function can be linked. Shepard has argued for the ubiquity of exponential decay functions for such data. Suitable substantive bases have yet to be formulated for other types of judgmental processes specific to methods of data collection.

Alternative Methods of Implementation

If the similarity between pairs of stimuli is to be estimated by an exponential decay function, then we have

$$S_{ij} = e^{-bD_{ij}} + c \tag{1}$$

where D_{ij} is the distance (Euclidean) between points in the stimulus space and b and c are the slope and asymptote, respectively, of the fitted exponential decay function.

Since this function is obviously monotonic, one might ask why we should be concerned with fitting it explicitly when Kruskal's [17] monotone regression algorithm will yield a reasonable step-like approximation, if indeed exponential decay is the best-fitting monotonic function. (It should be emphasized that nothing in the monotone regression algorithm predisposes the result to be a decay function.) The answer to this question may be seen by comparing Figures 1B and 1C. A detailed explanation and discussion of the stimulus space is given below, but for the moment we are concerned only with the schematic differences between stimulus points in the two different solutions obtained from the same matrix. (The contours representing clusters of stimuli are irrelevant to our argument and should be ignored for the present.) The solution in Figure 1B was obtained from a computer program designed by Chang and Shepard [6] that explicitly constrained the relationship between confusability and recovered distance (D_{ij}) to be an exponential decay function. Figure 1C represents the configuration from a straightforward application of Kruskal's [16, 17] MDSCAL. Arabie specified monotonic regression and obtained the solution in Figure 1C as having the minimum stress (.1199, stress formula one, primary approach to ties) from forty different random initial configurations used to circumvent local minima.

In comparing Figures 1B and 1C, it is apparent that there is considerable clumping of the stimuli in C. That is, stimuli that the subjects were able to discriminate with a nonzero probability are

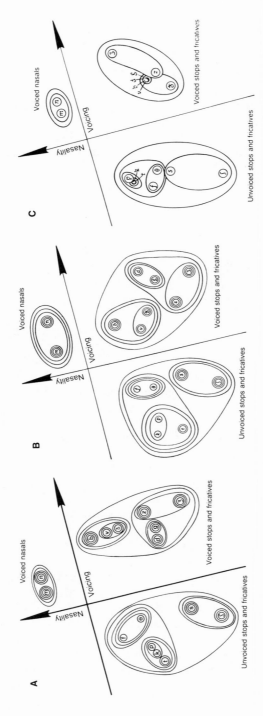

Figure 1. Two-dimensional spatial representations of sixteen consonant phonemes obtained from three multidimensional scaling analyses of the confusion measures of similarity collected by Miller and Nicely [21]. In A, the confusion measures were log transformed, and the regression function relating these values to interpoint distances was assumed to be linear. In B, untransformed confusion measures were used, with the regression function assumed to be exponential. In C, only a monotone relationship between the (untransformed) confusion measures and interpoint distances was assumed. (B and C are redrawn from Shepard [35].) The embedded hierarchical clusters (Shepard [34]) have also been included in each representation.

nonetheless placed in identical or nearly identical locations, as if the subjects always confused those stimuli. The stimuli separated by larger interpoint distances seem resistant to clumping in both solutions. Although the scatter plots of the regression functions are not presented here, they can be easily characterized. The function for B was (necessarily) the smooth exponential decay curve. In contrast, the scatter plot of the monotonic regression function of C had the appearance of being more like a step function, with plateaus separated by gaps.

The superior solution in Figure 1B (contrasted with C) was purchased with considerable effort, however. The program developed by Chang and Shepard [6] for explicitly fitting exponential decay functions is not generally available and has been used on very few data sets. The program [34, p. 74] is not the result of a simple modification of Kruskal's [16, 17] MDSCAL, since the former program requires (1) a rational initial configuration of the coordinates of the stimuli, (2) initial estimates of parameters b and c that afterward must be constrained to have appropriate signs, and (3) a gradient method that uses the second as well as the first partial derivatives of the goodness-of-fit function.

Given the difficulties of implementing the analysis leading to Figure 1B, it seems reasonable to seek an easier method for obtaining a solution devoid of the clumping produced by the straightforward nonmetric scaling in Figure 1C. The approach taken here begins with consideration of the exponential decay formula given above. If we assume that the asymptote c is approximately zero and take logs of both sides of the equation, we then have

$$\log S_{ij} = -bD_{ij} \qquad (2)$$

which specifies a linear regression function. Thus, logs of confusion probabilities can be analyzed simply by using the linear regression option in MDSCAL.

Two points should be emphasized. First, the log transformation causes the smallest confusion probabilities (which are likely to be the least reliable) to be differentially transformed into large distances. The fact that slight changes in those similarity values cause large fluctuations in distances (D_{ij}) has been cited as a difficulty for the type of analysis proposed here [29, p. 341; Shepard, personal communication]. Second, the log transformation as taken above results in an implicit transformation of Kruskal's [16, 17] badness-of-fit function, stress, that differs from the one in the

Chang and Shepard [6] program. The option allowed by MDSCAL for weighting the stimuli could be used to offset this difference, although we have not explored this possibility.

Figure 1A presents the scaling solution (.1787, stress formula one, primary approach to ties) obtained through the use of linear regression (fitting a slope and an intercept) on log-transformed data. It is apparent that most of the objectionable clumping found in C is not present in A. Moreover, the placement of stimulus points in A (linear regression) and B (Chang and Shepard's program) is very similar. (There will be a discussion below of the substantive nature of the two configurations.)

The present strategy of transforming the data for use of linear regression has several potential advantages over the approach of the Chang and Shepard [6] program. First, our goal is to find a prototypical form of regression for each of several methods of collecting proximities data. Our discussion of such forms has been given in atheoretical terms, however, the identification of prototypical functions would have important theoretical consequences. For instance, Shepard's [29] stochastic model for stimulus and response generalization represented an advance in psychological research on confusions in memory that is still useful years later [cf. 28], even though his model took as a primitive that exponential decay was the appropriate function. The search for prototypical functions specific to other types of tasks commonly used for collecting proximities could yield similar dividends of enhanced understanding of the types of judgment involved in such tasks. Moreover, our approach (viz., using the appropriate transformation to render the data tractable to linear regression) is clearly more desirable than requiring a specific algorithm and computer program for each method of proximity data collection having a specific rational regression function.

A second advantage appears when we move from two-way (viz., stimulus x stimulus) data sets, such as the one used as input for the solutions in Figure 1, on to three-way (viz., subjects or conditions x stimuli x stimuli) matrices. For the latter case, there is presently no program to fit decay functions explicitly, although such a program could be written. In fact, the most popular scaling program for three-way (or higher) matrices, Carroll and Chang's [3] INDSCAL, is a metric program whose nonmetric updates, NINDSCAL [5] and ALSCAL [37], are not yet as generally available. The approach suggested in the present paper would be to use the appropriate prototypical transformation before making routine use of a metric approach such as INDSCAL. Since nonmetric scaling methods are generally more expensive to use than their metric counterparts, a (transformed) metric analysis could achieve considerable economy for a large three-way data base. We now present an illustrative application of this approach to a well-known data set, considered as a two-way and later as a three-way matrix.

ILLUSTRATIVE APPLICATION OF THE APPROACH

The Miller and Nicely Data

The data used in the current analyses are from the classic Miller and Nicely [21] experiment on perceptual confusions between sixteen English consonant phonemes. Female subjects listened to female speakers read consonant-vowel syllables that were formed by pairing the consonants /b, d, g, p, t, k, v, ð, z, ʒ, f, θ, s, ʃ, m, n/ with the vowel /a/ (as in *father*), and the subjects were required to write down the consonant they heard after each syllable was spoken. Sound spectrograms of typical tokens of these syllables, from a female speaker, are shown in Figure 2. Miller and Nicely obtained full 16 x 16 matrices of confusions, or errors of identification, from the subjects' responses. The confusion matrices were collected under seventeen different experimental conditions, as listed in Table 1. These conditions fall under three headings: (1) noise-masking conditions varying speech-to-noise (S/N) ratios by adding random noise at different levels, (2) low-pass conditions[1] in which the only acoustic energy presented was that below a specified cutoff frequency, and (3) high-pass conditions in which the only acoustic energy presented was that above a specified cutoff frequency.

Results of the Two-way Analysis

Shepard's [34] analysis (discussed above; see also Figure 1B) used the matrix formed by pooling cell entries (see the appendix to this chapter) from the six so-called "flat-noise" conditions for which only the S/N ratio was varied. In our initial analysis, we took logs of the entries in Shepard's matrix[2] [34, p. 75] and used the option of linear regression in MDSCAL (with forty different random initial configurations) to obtain the solution in Figure A. It should be noted that the clusters of stimuli (grouped by closed curves) are from the complete-link dendrogram Shepard obtained from Johnson's [14] computer program for hierarchical clustering.

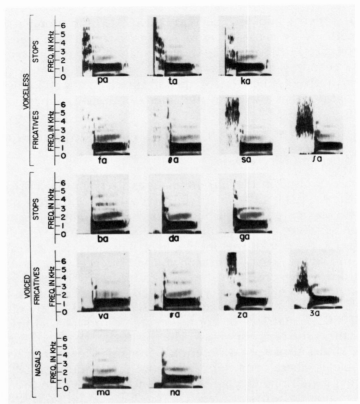

Figure 2. Sound spectrograms of typical utterances, by a female speaker, of the sixteen stimuli used in the Miller and Nicely [21] study of consonant confusability. Time runs from left to right (about a half-second per stimulus) in each spectrogram, while frequency in kHz increases from bottom to top. The consonant portion occupies approximately the leftmost third of each spectrogram. The consonant phonemes /θ/ and /ð/ are the *th* sounds in *th*in and *th*at, respectively, while /ʃ/ and /ʒ/ are the *sh* and *zh* sounds in *sh*ow and *Zh*ivago. (Figure and caption from Carroll and Wish [4, p. 77]; reproduced by permission of the authors and the publisher from *Contemporary Developments in Mathematical Psychology*, volume II, p. 77. Edited by Krantz et al. W. H. Freeman and Company. Copyright © 1974.)

Shepard [34] provided an elaborate interpretation of the clustering solution that pertains both to his data matrix and to our transformation of it, since complete-link cluster is monotone invariant. As clustering is not within the scope of this paper, we will not consider that form of data analysis and structural representation further.

TABLE 1
Dimension Weights from INDSCAL for the Seventeen Listening Conditions in Miller and Nicely's Experiment

Listening Conditions		Weights on Dimensions				R^2
Speech-to-Noise Ratio (dB)	Bandwidth (Hz)	Dimensions				(Proportion of Variance Accounted For)
		1	2	3	4	
Noise-masking Conditions:						
1. 12	200-6500	.35	.41	.43	.26	.58
2. 6	200-6500	.45	.53	.39	.24	.75
3. 0	200-6500	.52	.55	.37	.20	.81
4. -6	200-6500	.62	.52	.20	.17	.78
5. -12	200-6500	.74	.41	.20	.19	.84
6. -18	200-6500	.49	.40	.15	.10	.47
Low-Pass Filtering Conditions:						
7. 12	200-5000	.39	.54	.41	.26	.73
8. 12	200-2500	.45	.52	.43	.25	.77
9. 12	200-1200	.54	.52	.33	.10	.73
10. 12	200-600	.53	.59	.28	.13	.78
11. 12	200-400	.69	.41	.23	.16	.78
12. 12	200-300	.67	.50	.08	.05	.76
High-Pass Filtering Conditions:						
13. 12	1000-5000	.33	.37	.46	.38	.63
14. 12	2000-5000	.38	.15	.29	.55	.58
15. 12	2500-5000	.25	.21	.29	.56	.53
16. 12	3000-5000	.19	.10	.25	.69	.61
17. 12	4500-5000	.06	.08	.13	.77	.63
Proportion of Variance Accounted for by Each Dimension:		.33	.13	.16	.07	.69

The consonant configuration in Figure 1A closely resembles the configuration produced with the exponential regression function in B. Much of the same detailed structure can be observed in both configurations. Specifically, among the voiceless consonants, the stops /p, t, k/ are grouped closely together in both configurations, while the voiceless fricatives /f, θ, s, ʃ/ are ordered by place of articulation parallel to the vertical axis. In both solutions the voiced fricatives /v, ð, z, ʒ/ exhibit the same ordering as the voiceless fricative on the other side of the vertical axis. The voiced stops /b, d, g/ are not clearly separated from the fricatives in either configuration. Stops /d, g/ are to the left of /z/ in Figure 1A and to the right of /z/ in B. However, their psychological distances from the other voiced consonants in the current analysis are generally unaffected by this discrepancy. The configuration in A produced from the log-transformed matrix yields much the same information about the representation of the consonants in psychological space that the exponential analysis (Figure 1B) did, and both stand in marked contrast to the less informative configuration in C, produced with only monotonic constraints.

Results of the Three-way Analysis

The entire set of the seventeen Miller and Nicely confusion matrices were log transformed (see the appendix) for the three-way (conditions x stimuli x stimuli) scaling. This analysis used the Carroll and Chang [3] INDSCAL model and program. (The acronym stands for INdividual Differences SCALing, where "differences" in the present usage are between conditions in the experimental design, instead of between individual subjects.) Given the seventeen confusion matrices as input, this program generates an r-dimensional Euclidean stimulus space in which the consonant phonemes are embedded. These r dimensions are assumed to be relevant in each of the experimental conditions. However, each dimension may have differing degrees of perceptual importance for each condition, since there is a weight for every dimension-condition pair. That is, the estimated distance between a pair of phonemes, represented as j and k in this weighted Euclidean space, can be written

$$D_{jk}^{(i)} = \sqrt{\sum_{t=1}^{r} w_{it}(x_{jt} - x_{kt})^2} \tag{3}$$

where $D_{jk}^{(i)}$ is the distance between the stimuli j and k in the ith of the seventeen conditions, w_{it} is the weight for condition i on the tth dimension, and x_{jt} is the coordinate of stimulus j on the tth dimension.

Graphically, the output is represented as a stimulus space and a separate condition space. The former is represented as the usual display of stimuli in a coordinate space. However, unlike two-way nonmetric solutions in Euclidean space (e.g., Figure 1) with arbitrary orientations, the axes in an INDSCAL stimulus space have a mathematically preferred orientation. As such, the solution is only invariant over reflections and permutations of the axes. The data analyst, of course, is responsible for the substantive interpretation of these axes or dimensions.

The condition space is a plot of points representing the experimental conditions, with the fitted weights, w_{it} as coordinates, and it is usually in the positive orthant (although the INDSCAL program does not explicitly require the weights to be nonnegative). Psychologically, the weights gauge the importance or salience[3] of a dimension for each condition; statistically, the weights give an indication of the variance accounted for in data from a given condition. [See 3 for important details on normalization of the weights.]

Miller and Nicely's data have previously been analyzed with INDSCAL by Carroll and Wish [4, 41], although without any transformation of the matrices subsequent to symmetrizing the data. (See the appendix.) A comparison of the current analysis, also described in Soli and Arabie [36] with Carroll and Wish's six-dimensional solution should reveal whether the log transformation (which renders INDSCAL's linear regression model compatible with exponential decay) will improve the interpretability of the configuration and dimensions. The variances accounted for (VAF) by each dimension for the two analyses are compared in Table 2. Although the total VAF in six dimensions is approximately equal for both analyses, the VAF for the first four dimensions is consistently greater with the transformed data. The pattern of increases in VAF also differs. With the untransformed matrices, increments are about equal for each added dimension beyond the first, while increments for the transformed data were initially larger but drop rapidly after the fourth dimension is added. Therefore, the four-dimensional solution was selected, although its total VAF is slightly less than that for the six-dimensional solutions. The failure of VAF increments to level off by six dimensions with the untransformed data suggests that the fifth and sixth dimensions may have been needed to provide additional degrees of freedom for the linear model in fitting exponential data [18].

The solution for the sixteen consonant phonemes in the four-dimensional stimulus space is presented in Figure 3. (The horizontal

TABLE 2
Variances Accounted for (VAF) in INDSCAL Solutions for
Log-Transformed and Untransformed Confusion Matrices

Dimensionality	Transformed Matrices		Untransformed Matrices*	
	VAF	Increment	VAF	Increment
1	.33		.24	
		.13		.13
2	.46		.37	
		.16		.13
3	.62		.50	
		.07		.11
4	.69		.61	
		.04		.10
5	.73		.71	
		.04		.07
6	.77		.78	

*Data from Carroll and Wish [4].

and vertical broken lines segregate the consonants into the groups used for interpreting each of the dimensions.) The configuration in the first two dimensions resembles both Shepard's configuration in Figure 1B and Carroll and Wish's [4, p. 80] configuration in their voicing and nasality dimensions. However, the orientation of these first two dimensions (see Figure 3A) is quite different from those for the other solutions. This difference will be discussed in the summary after additional details of the current analysis have been presented.

Considering the first dimension, note that three clusters of consonants project onto it: the nasals /m, n/, the voiced stops and fricatives /b, d, g, v, ð, z, ʒ /, and the voiceless stops and fricatives /p, t, k, f, θ, s, ʃ/. These groups are positioned according to the temporal order of two acoustic events in the spoken syllable: the initial abrupt, burstlike release of acoustic energy and the onset of periodic pulsing. For the nasals, periodic pulsing, in the form of nasal resonance, precedes the burstlike release; with voiced consonants, the onset of periodic pulsing and abrupt release occur almost simultaneously; and, with voiceless consonants, pulsing begins after the release of acoustic energy. Thus, this dimension specifies the temporal order of the abrupt energy release and the onset of periodic pulsing and can be given the abbreviated label "periodicity/burst."

The projection of the consonant configuration on the second dimension (Figure 3A) also forms three groups. In the first group are

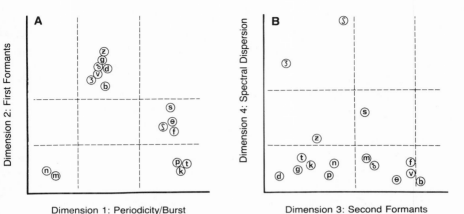

Figure 3. The four-dimensional "group" stimulus space from the INDSCAL analysis of log-transformed confusion measures of similarity between sixteen consonant phonemes obtained in seventeen different conditions [21]. The broken lines partition the planes into regions corresponding to distinguishable acoustic characteristics of the syllables.

the nasals and the voiceless stops, with voiced first formants that are relatively flat. In the second group are the voiceless fricatives that appear to have slightly rising voiced first formants. The voiced stops and fricatives form the third group at the other end of the dimension, with voiced first formants that clearly rise. This dimension seems to specify the shape of first formant transitions occurring in the presence of periodic pulsing and can be labeled "first formants."

The location of the consonants on the third dimension relates to their second formants' shape during periodic pulsing. (See Figure 3B.) This shape depends on both the consonant and the vowel in the syllable. In fact, a different ordering on this dimension would probably have been obtained had a different vowel been used [cf. 4]. Although the consonants are evenly distributed along the dimension, two groups may be distinguished for descriptive purposes. The first group contains those syllables with falling second formants during voicing and is best exemplified by /d, ʒ, g/, which have clearly observable falling transitions (Figure 2) and together occupy one end of the dimension. The second group, located in the middle of the dimension, consists of the syllables with voiced second formants that are relatively flat. The remaining phoneme, /b/, as the only consonant that has a rising second formant during voicing, occupies the other end of the dimension. Because this dimension separates the consonants on the basis of their second formant transitions during the voiced portion of the syllable, it can be given the label "second formants." Carroll and Wish [4] obtained a similar dimension in their analysis; however, the configuration of consonants was not quite as orderly, with /b/ failing to occupy the most extreme position opposite that of /d/ on the dimension.

The fourth dimension separates consonants into two small groups of fricatives, /ʒ, ʃ/ and /z, s/, and a large group comprising the other twelve phonemes. The fricatives /ʒ, ʃ/ exhibit large amounts of dispersion in their initial segments as shown in the spectrograms in Figure 2; /z, s/ show somewhat less dispersion; and the other twelve consonant phonemes contain almost none. Thus, the ordering of the consonants closely corresponds to the amount of dispersed spectral energy present in the initial part of each syllable; hence, the dimension is labeled "spectral dispersion." Carroll and Wish[4] reported that two dimensions, "sibilance" and "sibilant frequency," together separate these same fricative pairs from each other and from the remainder of the consonants. The need for *two* dimensions to segregate these pairs in the analysis with untransformed data, when a *single* dimension can perform the same function, may be a manifestation of the role of the extra fifth and sixth dimensions in fitting these data with a linear regression function.

So far, the interpretation of the four dimensions of the psychological space has been based on the projections of the consonant phonemes on each dimension. Interpretation of these dimensions may also be facilitated by examining their importance or salience for the listeners in the different experimental conditions. This information is contained in the configuration of points depicted by Figure 4. Noise-masking conditions are represented by the symbols N1 through N6 and the high- and low-pass conditions by H1 through H6, and L1 through L7, respectively. For each set of listening conditions, the degradation of the speech signal in each condition increases as the numbers increase.

The configurations of points representing the low-pass and noise-masking conditions on the first two dimensions are quite similar (Figure 4A). Noise masks the high frequencies more completely than low frequencies in speech, creating a perceptual effect similar to that of low-pass filtering [cf. 21]. The general pattern of

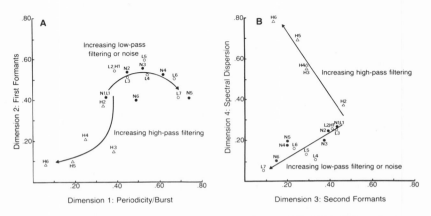

Figure 4. The four-dimensional "condition space" from INDSCAL. This analysis used log-transformed confusion measures of similarity between sixteen consonant phonemes obtained in seventeen different listening conditions [21]. Noise-masking, low-pass filtering, and high-pass filtering conditions are denoted by filled circles, open circles, and open triangles, respectively. Condition numbers increase with increasing degradation of the signal within each set of conditions. The arrows show the pattern of changing dimension weights within sets of listening conditions.

changes in salience with increasing degradation is indicated by the arrows superimposed on the configuration of points in the condition space. With increasing degradation, the relative salience of periodicity/burst and first formant information increases. However, in the most severe degradation conditions, the salience of the first formants returns to the original level observed for conditions N1 and L1. With severe degradation both dimensions are salient, suggesting that they are specified by relatively large amounts of acoustic energy in the lower portion of the spectrum. For most of the syllables, the abrupt release and the onset of voicing are, in fact, signaled with large amounts of acoustic energy in this spectral region (Figure 2). In contrast to the noise-masking and low-pass conditions, the pattern of weights is markedly different for high-pass conditions. As cutoff frequency increases from condition to condition, the salience of first formant transitions decreases drastically. However, the salience of periodicity/burst does not follow the same pattern. Acoustic energy from the burst is present throughout almost the entire spectrum, while periodic pulsing (i.e., the formants) is located only in the lower part of the spectrum. Since information for both the burst and the pulsing must be available if the periodicity/burst dimension is to be perceived, salience decreases only after high-pass cutoffs have excluded the pulsed portion of the signal.

The perceptual salience of second formant transitions and of spectral dispersion also exhibits similar changes with increasing degradation in low-pass and noise-masking conditions (Figure 4B). The spectral energy for these dimensions is generally less than that for the first two dimensions and is located in a higher frequency region of the spectrum. Consequently, when low-frequency bandwidth narrows or the S/N ratio decreases, the salience of both dimensions decreases. High-pass conditions again exhibit a pattern of weights very different from those in low-pass and noise conditions. With increasing cutoffs, the salience of second formant transitions decreases; however, at the same time, the importance of the spectral dispersion increases rapidly. Acoustic information for this dimension is dispersed throughout the high-frequency region and thus is the only perceptual information available under high-pass conditions with severe degradation.

The condition space does not convey all the information potentially available from the dimension weights. The several experimental conditions were systematically interrelated by the physical parameters specifying the kind and degree of acoustic degradation in each condition. Perceptual salience can be presented directly as a function of these parameters (viz., S/N, low-pass cutoff frequency, and high-pass cutoff frequency). Dimension weights will be presented in this format for only the high- and low-pass listening conditions, beginning with the formant transition dimensions. Weights for the first formant dimension are given in Figure 5A for both sets of conditions. In the experimental design the crucial difference between low-pass conditions was that low-pass cutoffs were varied while the high-pass cutoff was held constant at 200 Hz. The reverse was true for high-pass conditions, where high-pass cutoffs were varied while the low-pass cutoff was held constant at 5000 Hz. (See Table 1.) The variable cutoff frequencies for both low- and high-pass conditions are given on the horizontal axis of Figure 5A, with the condition number associated with each point also included. As we follow the low-pass function from right to left, changes in perceptual salience can be observed as the bandwidth in the lower portion of the spectrum narrows, thus increasing the degradation. Similarly, the salience for high-pass conditions can be read from left to right as the bandwidth in the upper portion of the spectrum narrows. When these functions are juxtaposed, the spectral region containing acoustic energy critical to each dimension

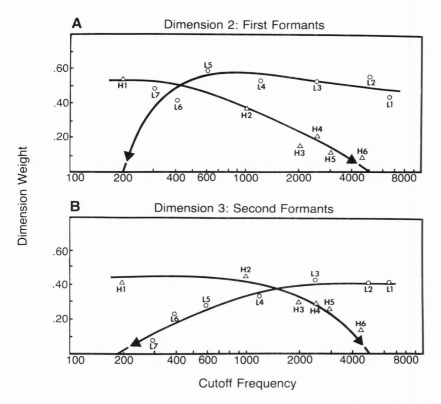

Figure 5. Dimension weights for the first- and second-formant dimensions in the high- and low-pass listening conditions. The weights are presented as a function of cutoff frequency, the crucial variable distinguishing within-group conditions. For the high-pass conditions, denoted by triangles, degradation increases with increasing high-pass cutoff frequencies as one moves in the direction of the arrow superimposed on the function, that is, from left to right in the figure. For low-pass conditions, denoted by circles, degradation increases with decreasing low-pass cutoff frequencies as one follows the function in the opposite direction, from right to left in the figure.

can be located. When salience first begins to decrease with additional decreases in signal bandwidth, a portion of this critical spectral region has been filtered out. We will define a region's boundary as the cutoff frequency at which salience begins to decrease. By such an examination of changes in weights with decreasing low-pass cutoffs, the upper boundary of the region can be ascertained. In the same way, the lower boundary can be found from the high-pass condition weights.

The location of the critical spectral region for the dimension labeled "first formant" can be seen to include that portion of the spectrum (approximately 200-700 Hz) in which these transitions usually occur. The functions for the perceptual salience of the second formant transitions, shown in Figure 5B, also clearly span the spectral region for this dimension. Salience begins to decrease for low-pass cutoffs less than 2500 Hz and for high-pass cutoffs greater than 1000 Hz, the region usually containing second formant transitions for female speakers.

The remaining dimensions, periodicity/burst and spectral dispersion (presented in Figures 6A and 6B, respectively), exhibit very different patterns of weights from those reported for the formant transitions. The high- and low-pass functions do not intersect; instead, the importance of periodicity/burst information in low-pass conditions is least with minimal degradation and increases consistently as the bandwidth narrows, suggesting that periodicity/burst information is perceptually important only when formant information is not available. In fact, for conditions with minimal degradation, both formant dimensions are perceptually more important than the periodicity/burst dimension. A similar phenomenon occurs with the spectral dispersion dimension. Its perceptual importance is greatest in the high-pass conditions with the most severe degradation; under relatively good listening conditions, spectral dispersion is the least important of the four dimensions. As degradation increased, so did the importance of the dimension to the listeners.

Summary of Analyses

Several points about these reanalyses of Miller and Nicely's [21] data can be made in summary. The two-way analysis of the transformed confusion matrices using the linear regression option in MDSCAL produced much the same information about the structure of the psychological space for the sixteen phonemes as Shepard's [34] analysis with the exponential regression function. The three-way analysis of the transformed matrices from all seventeen conditions indicated that additional informative structure was evinced when these transformed data satisfied INDSCAL's linear regression model. The parsimony of four dimensions favors the current solution over the earlier six-dimensional solution. Interpretation of these dimensions was explicated by convergent evidence from the configuration in the stimulus space, the configuration of dimension weights in the condition space, and the

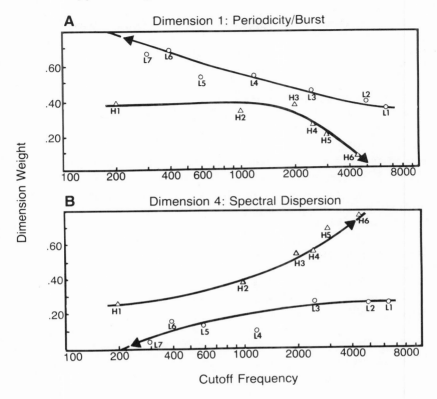

Figure 6. Dimension weights for the periodicity/burst and spectral dispersion dimensions in the high- and low-pass listening conditions. The weights are presented as a function of cutoff frequency, the crucial variable distinguishing within-group conditions. For the high pass conditions, denoted by triangles, degradation increases with increasing high-pass cutoff frequencies in the direction of the arrow superimposed on the function, that is, from left to right in the figure. For low-pass conditions, denoted by circles, degradation increases with decreasing low-pass cutoff frequencies in the opposite direction, from right to left in the figure.

systematic changes in the salience of each dimension as the band-pass region varied from condition to condition. Each dimension was well described in terms of the *acoustic* features of the signal (e.g., the temporal ordering of burst and periodicity). Masking noise and narrow bandpass filtering altered or eliminated portions of the acoustic information usually available to listeners, affecting the correspondence that would exist under normal listening conditions between acoustic cues and the phonetic features used to identify the consonants. The difference in correspondence can be seen by comparing the values of the weights for the four acoustic dimensions in a minimally degraded condition, the nearest approximation to normal listening conditions in this experiment, with the same values in a highly degraded condition. The relatively similar values that typify the former condition suggest one form of correspondence, and the divergent values within each degraded condition suggest another. Listeners in the latter conditions were forced to use the least-degraded acoustic information available to them when attempting to identify the syllables.

A similar phenomenon has been reported for comparisons of normal-hearing and hearing-impaired listeners in an experiment by Walden and Montgomery [39]. Normal-hearing listeners in that study, like the subjects in minimally degraded conditions of the Miller and Nicely study, weighted the consonant dimensions almost equally when making similarity judgments; hearing-impaired listeners, like subjects in degraded listening conditions, relied primarily on available acoustic information specified by one or two dimensions, again resulting in unequal dimension weights. Taken altogether, this evidence suggests that the dimensions of the psychological space for experiments in which the speech signal is altered, owing either to imposed degradation or hearing impairment, may well be defined in terms of the acoustic information available, rather than phonetic features.

The perceptual importance of periodicity/burst order and spectral dispersion only under conditions with severe degradation reinforces our preference for an acoustic interpretation of these dimensions. Neither of these sources of acoustic information supports a mapping between dimension(s) and traditionally emphasized feature(s); however, it is apparent that under listening conditions of severe degradation, those dimensions were crucial to correct identification. The evidence from the present INDSCAL analysis suggests that four acoustic dimensions best describe the psychological space for Miller and Nicely's experiment. The MDSCAL configurations in Figures 1A and 1B are in essential agreement with this assertion, although the arbitrary orientation of their dimensions is different. However, these axes may be rotated and reflected to correspond more closely to the first two acoustic dimensions, periodicity/burst and first formant transitions, yielding similar psychological spaces for each of the analyses.

APPENDIX: SYMMETRIZING THE MILLER
AND NICELY DATA

Among the various formulae that have been considered for symmetrizing the Miller and Nicely [21] data are

$$S_{ij} = \frac{s_{ij}}{s_{ii}} + \frac{s_{ji}}{s_{jj}} \qquad (A\text{-}1)$$

by Johnson [14],

$$S_{ij} = (s_{ij} + s_{ji})/2 \qquad (A\text{-}2)$$

by Wish and Carroll [41], and

$$S_{ij} = \frac{s_{ij} + s_{ji}}{s_{ii} + s_{jj}} \qquad (A\text{-}3)$$

by Shepard [34]. In formulae A-1 and A-2 the entries on the right-hand sides of the equations were summed across matrices from the six noise-masking conditions of the Miller and Nicely study. Among the objections raised by Hubert [12, p. 269] to the above formulae is the fact that the experimental design of Miller and Nicely resulted in unequal numbers of presentations for the different phonemes. Hubert, therefore, used the formula:

$$S_{ij} = s'_{ij} + s'_{ji} \qquad (A\text{-}4)$$

where s'_{ij} and s'_{ji} are normalized frequencies.

In addition to Hubert's caveats, other objections can be raised. For instance, formulas A-2 and A-4 discard the information contained in the diagonal entries. In spite of these potential misgivings about the approaches to symmetrizing the data, we adopted Shepard's (i.e., formula A-3) for reasons of comparability; like Shepard, we did not weight for unequal frequencies of occurrence.

In the three-way analyses, an additional problem was incurred, namely, the fact that 27.9% of the 2040 entries in the seventeen symmetrized confusion matrices were zeros. This perennial problem frequently arises in psychology when the log transformation is used (e.g., in taking the logs of magnitude estimates for stimuli at or near threshold in psychophysical tasks). From a methodological perspective, the problem obviously begs for a programmatic approach using parameter estimation of an additive constant.

However, in the present three-way analysis, we simply added 0.001 to each of the entries. This constant was intuitively selected in keeping with the absolute size of the entries. Some justification for the addition of 0.001 is provided by the fact that the product-moment correlation between $\log S_{ij}$ $(S_{ij} > 0)$ and $\log(S_{ij} + 0.001)$ was 0.9995.

A final detail that should be noted for the three-way INDSCAL analysis is that we specified the standard option for variances of the scalar products of the matrices from each experimental condition to be equalized, as was done by Wish and Carroll [41].

NOTES

1. Two of the seventeen conditions in Miller and Nicely's experiment can be considered as members of each of two of the groups of listening conditions. (See Table 1.) Specifically, the noise-masking condition with an S/N of 12 dB can also be viewed as the low-pass condition with the broadest bandwidth (200-6500 Hz), since that condition has the same high-pass cutoff and S/N as the other low-pass conditions. Similarly, the low-pass condition with a bandwidth of 200-5000 Hz, by having the same low-pass cutoff and S/N as the high-pass conditions, can be considered as the high-pass condition with the broadest bandwidth. When these two conditions are thus redundantly grouped, there is a total of six noise-masking conditions (N1-N6), seven low-pass conditions (L1-L7), where L1 and N1 are identical), and six high-pass conditions (H1-H6, where H1 and L2 are identical).

2. As noted earlier, in passing from formula 1 to formula 2, we are assuming that the asymptote, c for the exponential decay function is approximately zero. Shepard [34, p. 95] estimated that c equals 0.0003. The product-moment correlation between $\log S_{ij}$ and $\log(S_{ij} - 0.0003)$ is 1.0000 (rounded after more than four decimal places), where the S_{ij} are those listed by Shepard [34, p. 95] and used in both his and our two-way analyses. For a correction as large as $c = 0.001$, the correlation is 0.9999 but drops to 0.9953 for $c = 0.005$. Thus, assuming that c is zero appears to be inconsequential to the present analysis.

3. The weights represent the "strength of effect on perception" [3, p. 285] attributable to each of the dimensions in the seventeen listening conditions. A dimension can have a relatively high weight in a particular listening condition for two different reasons. First, in conditions with severely degraded speech signals, only dimensions that are resistant to that form of degradation will be salient and thus should have relatively large perceptual effects. For example, the periodicity/burst dimension was salient in extreme low-pass filtering and noise-masking conditions and consequently had a large perceptual effect. Second, a dimension that was not salient in highly degraded conditions can, nonetheless, have a relatively large perceptual effect in conditions with slight degradations. For example, the perceptual effect of the second formant dimension was greatest in such conditions. Evidently, this dimension was important and useful for identifying the consonant phonemes, although it was not salient. By way of contrast, the salient periodicity/burst dimension had less perceptual effect than the second formant dimension in the

conditions with slight degradations. Wang and Bilger [40] have reported a related finding. Using similar consonant identification tasks with various levels of masking noise, these authors found that in severe masking conditions the percentage of information transmitted, in terms of bits per stimulus, for the voicing and nasality features (which can be discriminated on the periodicity/burst dimension) was greater than that for any of the other fifteen features considered. However, in the absence of masking noise these two features no longer maintained their perceptual dominance over the other features. In order to avoid potential confusion about these differences in perceptual effects, dimensions will subsequently be designated as either perceptually *salient* or perceptually *important*, depending on whether they received their largest weights for highly degraded or slightly degraded conditions, respectively.

REFERENCES

1. Arabie, P., 1973. "Concerning Monte Carlo Evaluations of Nonmetric Multidimensional Scaling Algorithms," *Psychometrika*, 38, 607-8.
2. Arabie, P., S. M. Kosslyn, and K. E. Nelson, 1975. "A Multidimensional Scaling Study of Visual Memory in 5-Year-Olds and Adults," *Journal of Experimental Child Psychology*, 19, 327-45.
3. Carroll, J. D., and J.-J. Chang, 1970. "Analysis of Individual Differences in Multidimensional Scaling via an N-Way Generalization of Eckart-Young Decomposition," *Psychometrika*, 35, 283-19.
4. Carroll, J. D., and M. Wish, 1974. "Models and Methods for Three-Way Multidimensional Scaling." In D. H. Krantz, R. C. Atkinson, R. D. Luce, and P. Suppes (eds.), *Contemporary Developments in Mathematical Psychology* (vol. 2), pp. 57-105. San Francisco: W. H. Freeman.
5. Chang, J.-J., 1972. *Notes on NINDSCAL* (7/72 version). Mimeographed report, Bell Laboratories, Murray Hill, N.J.
6. Chang, J.-J., and R. N. Shepard, 1966. *Exponential Fitting in the Proximity Analysis of Confusion Matrices*, paper presented at the Annual Meeting of the Eastern Psychological Association, New York, April 14, 1966.
7. Cronbach, L. J., and G. C. Gleser, 1953. "Assessing Similarity between Profiles," *Psychological Bulletin*, 50, 456-73.
8. Ekman, G., 1963. "A Direct Method for Multidimensional Ratio Scaling," *Psychometrika*, 28, 33-41.
9. Farley, F. H., and A. Cohen, 1974. "Common-Item Effects and the Smallest Space Analysis of Structure," *Psychological Bulletin*, 81, 766-72.
10. Fisher, D. F., R. A. Monty, and S. Glucksberg, 1969. "Visual Confusion Matrices: Fact or Artifact?" *Journal of Psychology*, 71, 111-25.
11. Hovland, C. I., 1937. "The Generalization of Conditioned Responses: I. The Sensory Generalization of Conditioned Response with Varying Frequencies of Tones," *Journal of General Psychology*, 17, 125-48.
12. Hubert, L., 1972. "Some Extensions of Johnson's Hierarchical Clustering Algorithms," *Psychometrika*, 37, 261-74.
13. Indow, T., and T. Uchizono, 1960. "Multidimensional Mapping of Munsell Colors Varying in Hue and Chroma," *Journal of Experimental Psychology*, 59, 321-29.
14. Johnson, S. C., 1967. "Hierarchical Clustering Schemes," *Psychometrika*, 32, 241-54.

15. Keller, F. S., and R. E. Taubman, 1943. "Studies in International Morse Code II: Errors Made in Code Reception," *Journal of Applied Psychology*, 27, 504-9.
16. Kruskal, J. B., 1964. "Multidimensional Scaling by Optimizing Goodness of Fit to a Nonmetric Hypothesis," *Psychometrika*, 29, 1-27.
17. Kruskal, J. B., 1964. "Nonmetric Multidimensional Scaling: A Numerical Method," *Psychometrika*, 29, 28-42.
18. Kruskal, J. B., and R. N. Shepard, 1974. "A Nonmetric Variety of Linear Factor Analysis," *Psychometrika*, 39, 123-57.
19. Levelt, W. J. M., J. P. Van de Geer, and R. Plomp, 1966. "Triadic Comparisons of Musical Intervals," *British Journal of Mathematical and Statistical Psychology*, 19, 163-79.
20. Miller, G. A., 1969. "A Psychological Method to Investigate Verbal Concepts," *Journal of Mathematical Psychology*, 6, 169-91.
21. Miller, G. A., and P. E. Nicely, 1955. "An Analysis of Perceptual Confusions among Some English Consonants," *Journal of the Acoustical Society of America*, 27, 338-52.
22. Mostofsky, D. I. (ed.), 1965. *Stimulus Generalization*. Stanford, Calif.: Stanford University Press.
23. Plotkin, L., 1943. "Stimulus Generalization in Morse Code Learning," *Archives of Psychology* (New York), 40 (287).
24. Ramsay, J. O., 1968. "Economical Method of Analyzing Perceived Color Differences," *Journal of the Optical Society of America*, 58, 19-22.
25. Riley, D. A., 1968. *Discrimination Learning*. Boston: Allyn and Bacon.
26. Rothkopf, E. Z., 1957. "A Measure of Stimulus Similarity and Errors in Some Paired-Associate Learning Tasks," *Journal of Experimental Psychology*, 53, 94-101.
27. Rothkopf, E. Z., 1958. "Stimulus Similarity and Sequence of Stimulus Presentation in Paired-Associate Learning," *Journal of Experimental Psychology*, 56, 114-22.
28. Rumelhart, D. E., and A. A. Abrahamson, 1973. "A Model for Analogical Reasoning," *Cognitive Psychology*, 5, 1-28.
29. Shepard, R. N., 1957. "Stimulus and Response Generalization: A Stochastic Model Relating Generalization to Distance in Psychological Space," *Psychometrika*, 22, 325-45.
30. Shepard, R. N., 1958. "Stimulus and Response Generalization: Tests of a Model Relating Generalization to Distance in Psychological Space," *Journal of Experimental Psychology*, 55, 509-23.
31. Shepard, R. N., 1962. "Analysis of Proximities: Multidimensional Scaling with an Unknown Distance Function, I," *Psychometrika*, 27, 125-40.
32. Shepard, R. N., 1962. "Analysis of Proximities: Multidimensional Scaling with an Unknown Distance Function, II," *Psychometrika*, 27, 219-46.
33. Shepard, R. N., 1963. "Analysis of Proximities as a Technique for the Study of Information Processing in Man," *Human Factors*, 5, 33-48.
34. Shepard, R. N., 1972. "Psychological Representation of Speech Sounds." In E. E. David, Jr., and P. B. Denes (eds.), *Human Communication: A Unified View*, pp. 67-113. New York: McGraw-Hill.
35. Shepard, R. N., 1974. "Representation of Structure in Similarity Data: Problems and Prospects," *Psychometrika*, 39, 373-421.
36. Soli, S. D., and P. Arabie, 1979. "Auditory versus Phonetic Accounts of Observed Confusions between Consonant Phonemes," *Journal of the Acoustical Society of America*, 66, 46-59.
37. Takane, Y., F. W. Young, and J. de Leeuw, 1977. "Nonmetric Individual Differences

Multidimensional Scaling: An Alternating Least Squares Method with Optimal Scaling Features," *Psychometrika*, 42, 7-67.

38. Torgerson, W. S., 1958. *Theory and Methods of Scaling.* New York: John Wiley and Sons.
39. Walden, B. E., and A. A. Montgomery, 1975. "Dimensions of Consonant Perception in Normal and Hearing-Impaired Listeners," *Journal of Speech and Hearing Research*, 18, 444-55.
40. Wang, M. D., and R. C. Bilger, 1973. "Consonant Confusions in Noise: A Study of Perceptual Features," *Journal of the Acoustical Society of America*, 54, 1248-66.
41. Wish, M., and J. D. Carroll, 1974. "Applications of 'INDSCAL' to Studies of Human Perception and Judgment." In E. C. Carterette and M. P. Friedman (eds.), *Handbook of Perception* (vol. 2). New York: Academic Press.

Part 2
Preference Functions
and Choice Behavior

Chapter 2.1 Data Theory and Problems of Analysis

Lawrence Hubert

There is at least one distinct advantage to writing the first paper in a section devoted to a topic as broad as "preference functions and choice behavior" Compared to the more difficult task of providing the final summary and critique, an introductory statement for such an extensive subject allows a great degree of latitude in what can be emphasized. Granted, an overview that relates to the other papers in an integral manner is still expected, but a truly comprehensive analysis is not. This presentation takes full advantage of this sanction. I will limit myself to making a brief introduction to the problems of preference and choice based on a very specific literature in psychology and operations research that is most relevant to the study of paired comparison data. Although limited, the discussion will suggest one possible context in which the next four rather diverse selections could be read productively. Thus, my restriction has an organizational value that goes beyond merely introducing several analysis methods that may be unfamiliar to geographers.

As one very important informational item to help focus on appropriate evaluation strategies at the outset, I should note that all of the papers in this section bypass the type of joint unfolding

Partial support for this project was provided by the National Science Foundation, grant GSOC-77-28227.

analysis of individuals and objects popularized in the works of Coombs and his co-workers [4]. There is still an interest in identifying a latent continuum along which objects (for instance, stimuli) can be ordered, but this ordering and the generated preferences or assessments are assumed to be the same over all subjects. Stated differently, the unfolding problem of simultaneously placing individual subjects along the object continuum is not at issue, and no attempt is made to use a single dimension for generating different preference rankings or choices depending on the specific placement of an individual. The task of unidimensional scaling is still of paramount importance, but at least the behavioral data collected from individuals are not subject to unfolding and are assumed to be a direct indication of how a group of objects should be seriated.

The search for a unidimensional scale is common to the four papers that follow, but, depending on the specific topic addressed, this search takes on a different meaning in each. In the discussions of conjoint measurement (Burnett) and stimulus integration (Louviere), the unidimensional scale is based on combining object information according to certain rules. In the other two papers on revealed preference analysis (Clark) and spatial preference functions (Kohler and Rushton), the object ordering must be inferred from some sort of dominance function that is either given directly or generated according to a model. Since it is expected in all four cases that the resulting unidimensional scales could at some point be reflected behaviorally in paired comparison data, a single unifying context can be identified when we concentrate initially on the analysis of asymmetric proximity information. Each paper has obvious implications for evaluating such data or, at the very least, for predicting patterns that should be present when paired comparisons are obtained.

As one final introductory point, a basic issue could be raised regarding the source of models and statistical procedures that are typically used in studying preference and choice. Judging from the current literature in both psychology and geography, it appears that variants of regression and multidimensional scaling are of prime interest, which in turn places the associated problems of data analysis into the framework of model estimation. Although this latter perspective has traditionally formed the core of most popular data reduction strategies, a complete reliance on model estimation can also be rather restrictive, particularly when a substantive question can be phrased more directly as the verification of pattern in manifest data. In contrast, our interest will be in analysis strategies that are grounded in combinatorial optimization and nonparametric statistics. We do not intend these procedures to be replacements for those based on model estimation, but we do believe that they are valuable adjuncts that approach the topic of preference and choice from rather different first principles. Final substantive interpretations will tend to be consistent irrespective of the statistical procedures used, but it is always reassuring to know that data, rather than the chosen analysis methods, are responsible for the conclusions we would wish to make.

The selection of an analysis model that is both relevant to a given data set and directed toward the right substantive question is obviously a main concern of any researcher. In fact, the techniques presented throughout this book, as well as those in this paper, can all be interpreted in the context of locating the most appropriate model. Given the variety of data analysis schemes that are available in the literature, however, my brief presentation can in no way be seen as exhaustive of all the topics that could be developed even out of a specific operations research approach to the evaluation of preference and choice. The intent is more modest and merely points out one orientation that could be of value in the study of geographically important data sets. Parts of this orientation are already well developed in some areas of geography, although they are disguised in different forms. For example, the type of exploratory optimization heuristic illustrated later has direct parallels in the location-allocation literature [cf. 16]; the estimation of attractivities from migration statistics is really a version of unidimensional scaling using preference data [cf. 19]; and the confirmatory test I will use in comparing matrices relates to, and generalizes a portion of the spatial autocorrelation literature based on what Cliff and Ord [3] call the randomization model [10].

UNIDIMENSIONAL SERIATION

To introduce a convenient terminology, suppose we are given a set S of n objects, $\{O_1, O_2, \ldots, O_n\}$, and an asymmetric proximity function $q(\cdot, \cdot)$ defined on $S \times S$. The value assigned to the ordered pair (O_i, O_j), i.e., $q(O_i, O_j)$, will index the dominance of O_i over O_j; by convention, greater degrees of dominance will be reflected by larger values of the function. For example, the objects could refer to spatial locations and $q(O_i, O_j)$ could denote the proportion of times O_i was chosen over O_j when both were presented in a paired comparison choice task. As a technical convenience, it is assumed that $q(O_i,O_j) + q(O_j,O_i) = 1$. Thus, if m_{ij} denotes the raw number of times O_i was preferred to O_j, a natural measure of dominance could be defined very simply as $q(O_i,O_j) = m_{ij}/(m_{ij} + m_{ji})$.

The function $q(\cdot,\cdot)$ contains two distinct types of information that are relevant in sequencing the objects in S along a continuum. First, when appropriately seriated from left to right, the ordering of S should be reflected in the sign of $q(O_i,O_j) - q(O_j,O_i)$. Consequently, if the more dominant objects are conventionally placed to the left and O_i precedes O_j in this ordering, the expression $q(O_i,O_j) - q(O_j,O_i)$ should be positive. Second, it is expected that the absolute difference, $|q(O_i,O_j) - q(O_j,O_i)|$, would be directly reflected in the distance between the placements for O_i and O_j. No direction of dominance is now implicit, but large absolute differences should imply greater degrees of separation. In the foundational theory of measurement, this same sign and magnitude distinction is mirrored in the conditions for weak and strong stochastic transitivity [14].

The sign and magnitude information provided by differences of the form $q(O_i,O_j) - q(O_j,O_i)$ should lead to consistent results whenever a strong unidimensional scale underlies the original function $q(\cdot,\cdot)$. In fact, since the degree to which sign and magnitude data are compatible could be developed into a direct index of "scalability," it may be inadvisable in any analysis that has not demonstrated strong scale saliency either to eliminate one source from consideration altogether (for instance, signs are ignored in revealed preference analysis [15]) or to automatically combine both into a single proximity measure prior to showing that they would lead to similar substantive conclusions (as is done, for example, in direct metric parallels to revealed preference analysis that estimate the attractivities of spatial locations from migration statistics [19]).

As a very simple illustration of how these separate sources of information can be generated and used, the 18 x 18 matrix shown in Table 1 was selected from the paper by Clark (his Table 4). The matrix in this paper's Table 1 defines the values assigned by a function $q(\cdot,\cdot)$ for an object set S containing 18 members and forms only part of the 30 x 30 matrix analyzed by Clark. This reduction of 12 objects ensures that some of the later matrix comparisons in this paper are based on data sets that have no missing values. Multidimensional scaling, as used by Clark, handles missing proximities rather routinely, but for my procedures I would need to pursue a digression that is best delayed until the basic strategies are explained. Consequently, for pedagogical convenience, the spatial locations originally numbered 1, 6, 11, 12, 15, 20, 21, 24, 25, 28, 29, and 30 in Clark's discussion have been eliminated and the remaining locations sequentially relabeled by the integers 1 to 18. As a notation, the sign of $q(O_i,O_j) - q(O_j,O_i)$ when attached to unity will be denoted by $q^*(O_i,O_j)$. These values form the entries in the (skew-symmetric) matrix $\underset{\sim}{Q}^*$ given in the upper triangular portion of Table 2. In an analogous fashion, the absolute difference $|q(O_i,O_j) - q(O_j,O_i)|$ is denoted by $q^\Delta(O_i,O_j)$ and defines an

TABLE 1
An 18 x 18 Submatrix Taken from Clark's Table 4

	1	2	3	4	5	6	7	8	9	10	11	12	13	14	15	16	17	18
1	X	.53	.77	.81.	53	.72	.63	.40	.83	.74	.53	.84	.82	.68	.85	.46	.72	.81
2	.47	X	.78	.81	.40	.71	.68	.67	.82	.83	.47	.81	.83	.62	.81	.46	.67	.77
3	.23	.22	X	.60	.14	.43	.38	.36	.57	.49	.23	.56	.55	.30	.58	.19	.36	.52
4	.19	.19	.40	X	.06	.33	.32	.27	.45	.45	.10	.39	.55	.24	.43	.19	.16	.29
5	.47	.60	.86	.94	X	.71	.78	.90	.82	.88	.53	.82	.83	.83	.81	.47	.70	.80
6	.28	.29	.57	.67	.29	X	.49	.56	.66	.62	.30	.66	.66	.47	.64	.29	.49	.62
7	.38	.32	.63	.68	.22	.51	X	.43	.68	.62	.25	.67	.66	.48	.69	.26	.50	.66
8	.60	.33	.64	.73	.10	.44	.57	X	.60	.71	.20	.38	.70	.38	.61	.38	.33	.46
9	.17	.18	.43	.55	.18	.34	.32	.40	X	.43	.20	.48	.49	.31	.49	.16	.32	.46
10	.26	.17	.51	.55	.12	.38	.38	.29	.55	X	.20	.52	.53	.32	.54	.19	.36	.52
11	.47	.53	.77	.90	.47	.70	.75	.80	.80	.80	X	.80	.80	.72	.81	.47	.68	.80
12	.16	.19	.44	.61	.18	.34	.33	.63	.52	.48	.20	X	.50	.31	.51	.17	.33	.47
13	.18	.17	.45	.45	.17	.34	.34	.30	.51	.47	.20	.50	X	.31	.49	.14	.34	.48
14	.32	.38	.70	.76	.17	.53	.52	.62	.69	.67	.28	.69	.69	X	.72	.29	.50	.69
15	.15	.19	.42	.57	.19	.36	.31	.39	.51	.46	.19	.49	.51	.28	X	.17	.34	.47
16	.54	.54	.81	.81	.53	.71	.74	.62	.84	.81	.53	.83	.86	.71	.83	X	.73	.82
17	.28	.33	.64	.84	.30	.51	.50	.67	.68	.64	.32	.67	.66	.50	.66	.27	X	.65
18	.19	.22	.48	.71	.20	.38	.34	.54	.54	.48	.20	.53	.52	.31	.53	.17	.35	X

entry in the (symmetric) matrix Q^Δ given in the lower triangular portion of Table 2.

The task of unidimensional seriation can now be phrased by using either the Q^* or the Q^Δ matrix. In both instances, an object ordering is desired that would generate particular patterns in Q^* or Q^Δ when their rows and columns are rearranged. With Q^*, for example, a reordering of the rows (and simultaneously the columns) is desired that forces +1s above the main diagonal and −1s below.

TABLE 2
Sign Matrix Q^* Derived from Table 1 (above Diagonal)
Absolute Difference Matrix Q^Δ (below Diagonal)

	1	2	3	4	5	6	7	8	9	10	11	12	13	14	15	16	17	18
1	X	+1	+1	+1	+1	+1	+1	−1	+1	+1	+1	+1	+1	+1	+1	−1	+1	+1
2	.06	X	+1	+1	−1	+1	+1	+1	+1	+1	−1	+1	+1	+1	+1	−1	+1	+1
3	.54	.56	X	+1	−1	−1	−1	−1	+1	−1	−1	+1	+1	−1	+1	−1	−1	+1
4	.62	.62	.20	X	−1	−1	−1	−1	−1	−1	−1	−1	+1	−1	−1	−1	−1	−1
5	.06	.20	.72	.88	X	+1	+1	+1	+1	+1	+1	+1	+1	+1	+1	−1	+1	+1
6	.44	.42	.14	.34	.42	X	−1	+1	+1	+1	−1	+1	+1	−1	+1	−1	−1	+1
7	.26	.36	.24	.36	.56	.02	X	−1	+1	+1	−1	+1	+1	−1	+1	−1	0	+1
8	.20	.34	.28	.46	.80	.12	.14	X	+1	+1	−1	−1	+1	−1	+1	−1	−1	−1
9	.66	.64	.14	.10	.64	.32	.36	.20	X	−1	−1	−1	−1	−1	−1	−1	−1	−1
10	.48	.66	.02	.10	.76	.24	.24	.42	.14	X	−1	+1	+1	−1	+1	−1	−1	+1
11	.06	.06	.54	.80	.06	.40	.50	.60	.60	.60	X	+1	+1	+1	+1	−1	+1	+1
12	.68	.62	.12	.22	.64	.32	.34	.24	.04	.04	.60	X	0	−1	+1	−1	−1	−1
13	.64	.66	.10	.10	.66	.32	.32	.40	.02	.06	.60	.00	X	−1	−1	−1	−1	−1
14	.36	.24	.40	.52	.66	.06	.04	.24	.38	.36	.44	.38	.38	X	+1	−1	0	+1
15	.70	.62	.16	.14	.62	.28	.38	.22	.02	.08	.62	.02	.02	.44	X	−1	−1	−1
16	.08	.08	.62	.62	.06	.42	.48	.24	.68	.62	.06	.66	.72	.42	.66	X	+1	+1
17	.44	.34	.28	.68	.40	.02	.00	.34	.36	.28	.36	.34	.32	.00	.32	.46	X	+1
18	.62	.54	.04	.42	.60	.24	.32	.08	.08	.04	.60	.06	.04	.38	.06	.64	.30	X

The ordering that achieves the best pattern of this type defines a sequencing of the objects in S along a continuum—the more dominant objects are placed to the left [18]. Similarly, I can attempt to reorder the rows and columns of the matrix Q^* to approximate a perfect anti-Robinson condition, i.e., a pattern of entries that never decreases when moving within a row to the left or the right off the main diagonal [12]. Such an ideal sequencing exists when the function $q^*(\cdot,\cdot)$ is defined by a monotonic function of the (Euclidean) distances between the positions for the n objects placed along a single dimension. It should be noted that orderings based on Q^Δ are equivalent up to a complete reversal, since no orientation is automatically implied as it is for Q^*.

Combinatorial Optimization

The task of reorganizing the rows and columns of Q^* to force +1s above the main diagnonal and –1s below can be operationalized in terms of a simple index Γ defined by the sum of entries above the main diagonal. Although there is an enormous literature on this combinatorial optimization problem in psychology [8], some recent work on complexity in computer science suggests that computational heuristics will be necessary whenever n is greater than, say, 15. Up to this point and with reasonable computational costs, dynamic programming or branch-and-bound strategies are feasible; thus, there can be guarantees for an optimal solution. Since our example is over the limit, however, an efficient heuristic will be illustrated that is originally justified by several necessary conditions for an optimal solution discovered in operations research [20]. Operationally, we start with a (random) ordering of the rows and columns of Q^* and iteratively carry out pairwise interchanges of the rows (and columns) that maximize at each step the increase in Γ. Once a local maximum is achieved, a second stage is implemented in which arbitrarily sized subsets, consecutive in the given row (and column) order, are interchanged. After a local maximum is located in this second stage, the pairwise option is reimplemented, and so on, until a row (and column) order is identified that cannot be improved upon in either stage. Based on Q^* in Table 2 and 100 random starts for this two-stage procedure, two different index values were identified in the local maxima. A value of 138 was found 12 times and 140 was found 88 times. Although all 100 orderings were fairly consistent, minor differences were present even in those sequences that produced exactly the same value of Γ.

The ordering of 18 objects implied by Clark's one-dimensional solution is not a local maximum with respect to Γ, and, in fact, its index value of 112 was bettered for all 100 random initializations of the interchange procedure. Such a discrepancy is not unexpected, however, since different data analysis models are being used to obtain the seriations. Given the presence of three object pairs that were coded as zeros in Q^* because of identical values of $q(O_i,O_j)$ and $q(O_j,O_i)$, a maximum value of 150 would indicate a perfect order. No such ideal sequence was located in the local optima, but the number of reversals (entries below the main diagonal) for the "best" index value of 140 (i.e., 10 reversals) was still less than the 38 reversals present for the sequence induced by Clark's scaling solution. This latter comparison is somewhat unfair, given that I have extracted a subsequence from Clark's original scaling solution and actually optimized Γ to attain the value of 140. As shown below, however, I would still do better in terms of Γ when a seriation is constructed from the data provided by Q^Δ.

The symmetric matrix Q^Δ can be analyzed in a similar manner by using a different objective function. Although many different possibilities exist for defining an appropriate index, only one of the very simplest will be considered as an illustration. Suppose I attempt to reorder the rows (and columns) of Q^Δ by placing the 18 objects into 18 positions spaced one unit apart. The adequacy of a placement will be indexed by a measure Λ defined by the Pearson correlation between the values assigned by $q^\Delta(\cdot,\cdot)$ and the reconstructed distance measures along the equally spaced continuum. Again, starting with 100 random orderings of Q^Δ, a two-stage heuristic was applied. A pairwise interchange defined the first stage; the second stage permitted the insertion of any consecutive sub-sequence between any other object pair in the given order. Only one local maximum, with a correlational value of .844, was identified (5, 16, 11, 1, 2, 14, 17, 7, 6, 8, 3, 18, 10, 12, 15, 13, 9, 4). The ordering suggested by Clark's solution (5, 16, 11, 1, 2, 6, 7, 17, 14, 3, 8, 13, 10, 12, 18, 9, 15, 4) is not a local optimum with respect to this correlational index and has a value of .774. Obviously, the Clark solution was bettered by all 100 random initializations, but it may be more significant to note that the best ordering based on Q^Δ had a smaller number of reversals than Clark's sequence when used on the sign matrix Q^* (a Γ index of 130 compared to 112 for the scaling solution).

Since all of the best orderings of S are more or less consistent when based on multidimensional scaling or combinatorial optimization on Q^* or Q^Δ, a unidimensional representation is probably appropriate for the Table 1 data. The minor variations that do appear, however, emphasize the difficulty of specifying the "fine" structure of any object seriation based on fallible data, particularly when different models are being used. Gross details of the orderings may be similar, but the choice of the optimization criterion strongly influences the more subtle aspects of the seriation.

Given the "best" orderings achieved for Q^* and Q^Δ through combinatorial optimization, there is an explicit cross-validation strategy that could be followed. If the original data represent a unidimensional scale as conjectured, the local maximum in each case (up to complete reversals in the orderings obtained for Q^*) should be more or less consistent. In fact, if we are committed to a nonmetric multidimensional scaling approach, it may be of value to use the sequences obtained from Q^* and Q^Δ as initial configurations. This last comment is especially pertinent given the troublesome problem of local optima when nonmetric multidimensional scaling in one dimension is used. Quoting from Shepard [17, pp. 378-79] on this anomaly:

> Ironically, local minima pose especially prevalent and therefore irksome obstacles to the attainment of the optimum configuration in what might otherwise seem to be the simplest case; viz., that of a one-dimensional space. Theoretical analysis confirms what has become clear from practical experience. Evidently, a point that is initially situated on the wrong side of some other points can gradually work its way around those other points in a space of two or more dimensions but, when confined to a single line, is unable to move through those points owing to forces of mutual repulsion. Even in published studies, one-dimensional solutions have been presented that I can show to correspond to such merely local minima.

Since this problem of "working past" objects does not occur to the same degree in the simple optimization heuristics I have used on Q^* and Q^Δ, it could be argued that single-dimension solutions are best approached by strategies other than nonmetric multidimensional scaling. At the very least, we may wish to follow Kendall's [12] lead and seek single-dimension solutions in two-dimensional representations.

Before I leave the topic of combinatorial optimization, several additional points should at least be mentioned in passing that would indicate the larger context in which both of the reordering problems discussed above can be placed. For convenience, I will merely state these ideas in summary form and provide references to more thorough discussions.

First of all, the optimization criteria that have been discussed for Q^* and Q^Δ are easily generalized to indexes that depend only on the rank order of the entries in a matrix [9]. In this way, the criticism that my indexes are not as general as those used in non-metric multidimensional scaling can be answered. Moreover, if I were convinced that a strong unidimensional scale were present and wished to identify a good sequencing along a continuum using the original asymmetric data function $q(\cdot,\cdot)$ without splitting it into two parts, an extensive theoretical justification is available in the form of maximum likelihood paired comparison ranking [2]. Here, a measure such as Γ would be used that was specified as the sum of above-diagonal proximities. A similar literature on the topic of quadratic assignment deals with maximizing measures such as Λ [13].

Locating unidimensional scales through combinatorial optimization is one of the most well developed topics in operations research and related areas of psychology. Readers interested in pursuing the more current work in the area are referred to Defays [5], deLeeuw and Heiser [6], and Baker and Hubert [1]. This latter reference includes various generalizations of the optimization strategies discussed above that includes a method for locating *subsets* of S that are orderable along a single dimension. Also, it is possible to extend all of these ideas to sets of K matrices by using a theory of generalized concordance [11].

CONFIRMATORY TESTING

The topic of combinatorial optimization has obvious significance whenever one wishes to identify a structure underlying a preference function. There is an equally important second analysis area, however, that is more concerned with the verification of conjectured pattern. For example, when each object in the set S has an associated variable (such as density of housing, income, and so on), it may be appropriate to formulate a direct test of whether the preference function mirrors the information contained within these supplementary data. Or, possibly, a model such as that proposed in stimulus integration theory may suggest certain properties of a proximity function, and the task is to assess the adequacy of these predictions. In fact, predictions can even be obtained empirically from a second proximity function, in which case an assessment of similarity across two samples would be obtained.

In all of these illustrations, the inference problem can be reduced to one of comparing two matrices and evaluating the degree of observed correspondence. Fortunately, there is a large literature on this task that can be applied directly once the necessary comparisons are reformulated in an appropriate manner. To start with some general notation, it is assumed that two proximity functions are available on $S \times S$, denoted by $q_1(\cdot,\cdot)$ and $q_2(\cdot,\cdot)$. The degree of relationship between the values assigned by these two functions can be measured in many different ways [9], but a very simple starting point is the well-known Pearson product-moment correlation. Considering the values of $q_1(\cdot,\cdot)$ and $q_2(\cdot,\cdot)$ over all $n(n-1)$ distinct object pairs, the Pearson index may be an obvious descriptive statistic even though the statistical inference scheme that is typically followed would ignore any algebraic linkages among the proximity values. For this reason, several investigators have argued that a more reasonable "null" model should be assumed, explicitly, one that would construct a reference distribution for the Pearson measure under the hypothesis that all $n!$ possible relabelings of the rows and simultaneously the columns of one of the proximity matrices are equally likely (based equivalently on either $q_1(\cdot,\cdot)$ or $q_2(\cdot,\cdot)$). This topic is argued in detail elsewhere, and, consequently, the reader is referred to Hubert [9]. For present purposes, I will give one very simple example based on these ideas and then note several interpretations for the two functions $q_1(\cdot,\cdot)$ and $q_2(\cdot,\cdot)$ that relate directly to the four papers in this section.

Table 3 presents an analogue to Table 1 based on a second sample presented in Clark's paper (his Table 5). Relying on the Pearson index mentioned above, a rather unremarkable correlation of .527 was obtained between Tables 1 and 3. Although I could assert significance of this value at an approximate level of .02 by using the type of Monte Carlo testing strategy discussed in Cliff and Ord [3] and elsewhere (based on a random sample size of 99 from all 18! reorderings of one of the matrices), the 28% shared variance (as specified by the square of .527) is not that impressive. It could be expected, therefore, that scaling results based on Tables 1 and 3 may differ markedly. For example, as reflected in Clark's nonmetric multidimensional scaling analysis, the orderings induced on the 18-object set (using the complete matrices from which Tables

1 and 3 were obtained) have only a nonsignificant tau value of .16. Given the low percentage of variance shared by the raw data matrices, this last result is not surprising and would indicate that a clear interpretation for orderings based on Tables 1 and 3 may be very difficult to obtain.

Considering the example illustrated above, it is apparent that the work of Burnett and Louviere is at a much more basic and detailed level. Nevertheless, the same approach to matrix comparisons may still help in evaluating the specific algebraic model-building strategies implicit in the theories of conjoint measurement and stimulus integration. Given the prediction of paired comparison choices (and possibly their magnitudes) from the constructed unidimensional scales based on averaging, additivity, multiplicativity, and so on, these predictions can be tested on a second paired comparison matrix obtained for a different sample of subjects. The first matrix defined from $q_1(\cdot,\cdot)$ now represents the empirically obtained choice proportions and the second function $q_2(\cdot,\cdot)$ could be defined, say, by the absolute differences between the previously derived scale values. Or, possibly, when the basic forms of, say, the stimulus integration model (e.g., additivity versus multiplicativity) lead immediately to contrary paired comparison predictions, a correlational index could be obtained for each subject against each set of predictions and used as a dependent measure in a traditional repeated measures design. Thus, an a priori test of which model would be the more appropriate could be reformulated in terms of an analysis of variance problem.

CONCLUDING COMMENTS

Because of missing data, my examples based on Tables 1 and 3 were restricted to an object set smaller than Clark's. Since multidimensional scaling can routinely handle missing data, it is important to point out how my analysis strategies can be modified to include incomplete data matrices when they are to be considered viable alternatives to scaling. First of all, in all the exploratory and confirmatory problems considered, the relevant statistics are equivalent (up to a fixed normalization) to a cross-product measure between two proximity functions $q_1(\cdot,\cdot)$ and $q_2(\cdot,\cdot)$, i.e., $\sum_i \sum_j q_1(O_i,O_j)q_2(O_i,O_j)$. For example, in defining Γ from the sign information in $q(O_i,O_j) - q(O_j,O_i)$, the first function $q_1(\cdot,\cdot)$ corresponds to $q^*(O_i,O_j)$ and $q_2(\cdot,\cdot)$ is an upper triangular matrix of all 1s (0s below the main diagonal). Similarly, if $q^\Delta(\cdot,\cdot)$ is interpreted as $q_1(\cdot,\cdot)$, then $q_2(O_i,O_j)$

TABLE 3

An 18 x 18 Submatrix Taken from Clark's Table 5

	1	2	3	4	5	6	7	8	9	10	11	12	13	14	15	16	17	18
1	X	1.00	.81	1.00	.43	.75	1.00	1.00	.80	1.00	.25	.73	.82	1.00	1.00	.56	.00	1.00
2	.00	X	1.00	1.00	.00	.50	.75	1.00	.86	.90	.00	.67	.75	.78	.43	.40	.00	.29
3	.19	.00	X	.00	.07	.13	.25	1.00	.18	.24	.06	.13	.13	.18	.19	.00	.00	.00
4	.00	.00	.00	X	.00	.00	.00	.30	.05	.05	.00	.05	.00	.10	.13	.04	.00	.00
5	.57	1.00	.93	1.00	X	.89	1.00	1.00	.94	.94	.25	.83	.87	1.00	.80	.67	.25	.73
6	.25	.50	.87	1.00	.11	X	.80	1.00	.50	.88	.00	.33	.57	.89	.33	.17	.00	.09
7	.00	.25	.75	1.00	.00	.20	X	1.00	.25	.57	.00	.14	.00	.57	.33	.11	.00	.00
8	.00	.00	.00	.70	.00	.00	.00	X	.00	.20	.00	.20	.00	.33	.43	.17	.00	.00
9	.20	.14	.81	.95	.06	.50	.75	1.00	X	.74	.06	.21	.50	.82	.22	.24	.06	.06
10	.00	.10	.76	.95	.06	.13	.43	.80	.26	X	.00	.22	.27	.40	.19	.07	.25	.24
11	.25	1.00	.94	1.00	.75	1.00	1.00	1.00	1.00	.94	X	1.00	1.00	1.00	1.00	.83	.00	.91
12	.27	.33	.88	.95	.17	.67	.86	.80	.79	.78	.00	X	.85	.82	.46	.31	.00	.17
13	.18	.25	.88	1.00	.13	.43	1.00	1.00	.50	.73	.00	.15	X	.90	.25	.11	.00	.07
14	.00	.22	.82	.90	.00	.11	.43	.67	.18	.60	.00	.18	.10	X	.36	.10	.00	.00
15	.00	.57	.81	.88	.20	.67	.67	.57	.78	.81	.17	.54	.75	.64	X	.30	.00	.42
16	.44	.60	1.00	.96	.33	.83	.89	.83	.76	.93	.75	.69	.89	.90	.70	X	.17	.55
17	1.00	1.00	1.00	1.00	.75	1.00	1.00	1.00	1.00	.94	.00	1.00	1.00	1.00	1.00	.83	X	1.00
18	.00	.71	1.00	1.00	.07	.01	1.00	1.00	.04	.76	.00	.82	.93	1.00	.58	.45	.00	X

could be defined as $|i-j|$. Consequently, when missing data are an issue, the simplest modification that might be considered is a restriction of the double sum to pairs of indexes for which proximity entries exist in both $q_1(\cdot,\cdot)$ and $q_2(\cdot,\cdot)$.

Since the number of defined terms in a cross-product may differ when the various statistics are obtained in constructing a reference distribution or in carrying out a heuristic optimization, it may be appropriate to normalize each raw cross-product sum by the number of terms available. In other words, since $n(n-1)$ terms are used when data are complete, the cross-product sum would be divided by a number smaller than $n(n-1)$, depending on how many expressions of the form $q_1(O_i,O_j)q_2(O_i,O_j)$ have at least one missing factor. When this normalization is carried out during the construction of a reference distribution or in the course of optimizing the raw cross-product sum, the divisors may vary from statistic to statistic, depending on which entries in $q_1(\cdot,\cdot)$ amd $q_2(\cdot,\cdot)$ are multiplied together.

Generalizations of the type illustrated above for missing data obviously extend the range of application for the evaluation methods we have emphasized. This is only a beginning, however, and many other variations and extensions are possible. For example, as mentioned at the beginning of this paper the work on spatial autocorrelation reflects one obvious research direction, as does the extensive literature in location theory, particularly when standard objects, i.e., "fixed locations," are included in an analysis of preference and choice [cf. 7]. Consequently, besides looking for work outside geography that is relevant to the analysis of preference and choice, it may be just as appropriate for geographers to search within their own discipline for various data analysis models that may be just as powerful.

REFERENCES

1. Baker, F. B., and L. J. Hubert, 1977. "Applications of Combinatorial Programming to Data Analysis: Seriation Using Asymmetric Proximity Measures," *British Journal of Mathematical and Statistical Psychology*, 30, 154-64.
2. Bowman, V. J., and C. S. Colantoni, 1973. "Majority Rule under Transitivity Constraints," *Management Science*, 19, 1029-41.
3. Cliff, A. D., and J. K. Ord, 1979. *Spatial Autocorrelation*. London: Pion.
4. Coombs, C. H., 1964. *A Theory of Data*. New York: John Wiley and Sons.
5. Defays, D., 1978. "A Short Note on a Method of Seriation," *British Journal of Mathematical and Statistical Psychology*, 31, 49-53.

6. deLeeuw, J., and W. Heiser, 1977. "Convergence of Correction-Matrix Algorithms for Multidimensional Scaling." In J. C. Lingoes (ed.), *Geometric Representations of Relational Data,* pp. 735-52. Ann Arbor, Mich.: Mathesis Press.

7. Francis, R. L., and J. A. White, 1974. *Facility Layout and Location.* Englewood Cliffs, N.J.: Prentice-Hall.

8. Hubert, L. J., 1976. "Seriation Using Asymmetric Proximity Measures," *British Journal of Mathematical and Statistical Psychology*, 29, 32-52.

9. Hubert, L. J., 1978. "Generalized Proximity Function Comparisons," *British Journal of Mathematical and Statistical Psychology*, 31, 179-92.

10. Hubert, L. J. 1978. "Nonparametric Tests for Pattern in Geographic Variation: Possible Generalizations," *Geographical Analysis*, 10, 86-88.

11. Hubert, L. J., 1979. "Generalized Concordance," *Psychometrika*, 45, 135-42.

12. Kendall, D. G., 1971. "Seriation from Abundance Matrices." In F. R. Hodson et al. (eds.), *Mathematics in the Archaeological and Historical Sciences*, pp. 213-52. Edinburgh: Edinburgh University Press.

13. Lawler, E. L., 1975. "The Quadratic Assignment Problem: A Brief Review." In B. Roy (ed.), *Combinatorial Programming: Methods and Applications.* Dordrecht, Holland: D. Reidel, 351-60.

14. Luce, R. D., and P. Suppes, 1965. "Preference, Utility, and Subjective Probability." In R. D. Luce et al. (eds), *Handbook of Mathematical Psychology* (vol. 3), pp. 249-41. New York: John Wiley and Sons.

15. Rushton, G., 1969. "The Scaling of Locational Preferences." In K. R. Cox and R. G. Golledge (eds.), *Behavioral Problems in Geography: A Symposium.* Evanston, Ill.: Department of Geography, Northwestern University, Studies in Geography #17.

16. Scott, A. J., 1971. *Combinatorial Programming: Spatial Analysis and Planning.* London: Methuen.

17. Shepard, R. N., 1974. "Representation of Structure in Similarity Data: Problems and Prospects," *Psychometrika*, 39, 373-421.

18. Slater, P., 1961. "Inconsistencies in a Schedule of Paired Comparisons," *Biometrika*, 48, 303-12.

19. Tobler, W. R., 1979. "Estimation of Attractivities from Interactions," *Environment and Planning A*, 11, 121-27.

20. Younger, D. H., 1963. "Minimum Feedback Arc Sets for a Directed Graph," *IEEE Transactions on Circuit Theory*, 10, 238-45.

Chapter 2.2 Recovering the Parameters of a Decision Model from Simulated Spatial Choice Data

James A. Kohler and *Gerard Rushton*

Geographers have long been interested in how people choose a place for a particular purpose from among the many places available. Early modeling efforts had as their goal the replication of observed flows. These models were various forms of the gravity model and most often were transformed into linear regression models for solution by least squares methods. Correlation coefficients were generally regarded as adequate measures of goodness of fit, and the regression coefficients were interpreted causally as reflecting the degree of importance of the variable in question. Thus, large coefficients on the variable distance were interpreted to mean that distance strongly influenced the choice of destination. More recently, however, it became clear that the recovered parameters bore no known relationship to any model of choice because their values were influenced as much by the distribution of alternatives as by the choice model [2, 5, 8]. When gravity-type models of spatial interaction were used to model spatial choice data generated by identical choice models applied to different environments, the recovered parameters were different [1, 9, 10]. Hence, it was clear that such models could not be used to identify the spatial choice model used in any empirical situation.

This marked variability of parameter values in empirical studies ushered in a model-building period in which the emphasis was no

longer exclusively on constructing models that accurately repli-
cated the observed choices but rather was centered on the construction of models in which both the functional form and the
parameter values could be interpreted. Behavioral, decision models
of choice appeared to many to belong to this class. Choice models,
it is claimed, are valid not only in the sense that they correctly
describe a large proportion of the spatial choices observed, but
also in the sense that they are effective in a wide variety of conditions. Two choice models have addressed this problem. They are
the preferred choice model constructed through revealed preference analysis on observed choices and the information-integration
model constructed through the responses of subjects to hypothetical situations usually reflecting combinations of relevant attributes
of the choice situation [3, 6].

Both approaches claim to be methodologies that deal with how
people choose between the fundamental attributes that influence
the spatial choice decision and with the rule of combination (or
function) by which attributes are weighted to give a net degree of
attraction to each opportunity and thus the prediction of choice.

The difference between the two approaches is that information-
integration approaches have the advantage of presenting many
combinations of attributes to people in a systematic fashion. Such
experimental designs generate a great deal of data and allow the
experimenter to compute mathematical functions showing how
the attributes are combined. Revealed preference approaches have
an advantage in that they work with data on choices that are actually observed. They have a disadvantage in that a person making a
spatial choice betrays only a small part of his or her information-
processing capability. Any given choice selected can be shown to
be consistent with a wide variety of information-processing functions, so that, as far as any one individual is concerned, ambiguity
is likely to remain after his or her choices have been analyzed.

The disadvantage of integration-theory approaches is that they
show how people combine those attributes that are presented to
them in experimental situations. It is only recently that studies
have been designed to validate the transfer of these models to empirical spatial choice situations.

We regard these two methodologies as complementary to the
task of analyzing spatial choices to derive models that are robust
with respect to a variety of observed situations. They are complementary because there are situations in which only observed
choices are available from which, in the short run, models are to

be constructed. There are also situations in which choices are not available, and only choices between hypothetical alternatives are available from which generalization is needed. Then there is a common ground where actual choices exist and where it is also possible to generate choices between hypothetical alternatives in an information-processing context. In such situations, it should be possible to use the revealed preference data from observed choices to validate the function developed from the integration-theory approach. When validated, such a function often investigates the choice model in a wider variety of contexts than is possible when observed choices are used.

It is interesting that the problems the two methodologies present us with at this time are quite different. In the case of functional measurement models or information-integration models, it is possible to recover the function and its parameters by which subjects have combined the attributes with which they have been presented. But the problem is in establishing that this function is actually used in empirical situations and that it is not confounded by other variables that were not included in the experimental situation.

In the case of revealed spatial preference research, it seems reasonable to claim that, if the weighting function by which attributes were combined were found, it would probably be useful in predicting observed spatial choices. So, in the first case, finding the function is not much of a problem but establishing its generality is; in the second case, the generality of the function seems assured if it can, in fact, be found from revealed preference data.

OBJECTIVES

In this paper we investigate the degree to which the parameters of a spatial choice model can be accurately recovered through the knowledge of observed spatial choices and of the alternatives that were rejected. We also investigate the relationship between sample size and parameter recovery. In judging the goodness of fit of the procedure, we also find it necessary to develop criteria for measuring the accuracy of parameter recovery.

METHODS

Our approach is to generate spatial choices in a simulation model using a hypothetical information-processing function. We then accept these spatial choices and, without using the knowledge of the function parameters that generated them, attempt to recover the parameters.

RECOVERING THE MODEL PARAMETERS

Our method for finding the parameter values involves first recovering the preference scale and then transforming it by use of an iterative algorithm so that preferential distances between stimuli are accurately recovered. Finally, we compute the parameters of the function by a least squares fit to the transformed preference scale values.

Recovering the Preference Scale

To find the preference scale we developed a matrix of paired comparison probabilities from the simulated spatial choice data showing the probability that places at a given combination of distance and site attraction were chosen over places at some other combination of distance and site attraction when both combinations were present and one was chosen. The methodology for constructing a scale from such a matrix has been described elsewhere [6]. It involves the application of a nonmetric multidimensional scaling algorithm on an interpoint distance matrix defined as:

$$s_{ij} = |p_{ij} - 0.5| \qquad\qquad 0 \leqslant s_{ij} \leqslant 0.5$$

where:

$$p_{ij} = \frac{x_{ij}}{x_{ij} + x_{ii}}$$

and where x_{ij} equals the number of times ith stimuli combination is chosen when both ith and jth stimuli are present. Similarly, s_{ji} relates to choice of jth stimuli combination in the presence of the ith combination.

Transforming the Scale

If a scale (s_i) has been computed by nonmetric scaling techniques from the s_{ij}, then distances computed from that scale (d_{ij}) should increase monotonically as the original s_{ij} values increase, providing, of course, that an underlying scale exists. The objective in transforming the scale is to find new scale values such that distances computed between all points of the scale are equal to distances between equivalent points on the original scale that was used in simulating the spatial choice data. The function we used to simulate the spatial choice data was:

$$z_i = P_i^\alpha / D_i^\beta$$

where P_i is the place attraction value and D_i the distance to the place ($\alpha = 0.8$, $\beta = 2$).

Dr. Ming-Mei Wang has shown us that a scale computed by non-metric multidimensional scaling from the s_{ij} values generated directly from this scale will have regression weights that are some multiple of the original parameter value when the recovered scale values are regressed against the logarithms of the variables that characterize the stimuli combinations. (See the appendix for Dr. Wang's proof.) That is:

$$x_i = \theta \alpha \ln P_i - \theta \beta \ln D_i + b$$

where θ is the unknown factor.

Algorithm for Recovering Parameters

With this information, a two-stage algorithm was constructed for recovering the preference function parameters. The first stage involves performing a linear regression of the following form:

$$Y = b_1 \ln P_i + b_2 \ln D_i + a$$

where P_1 and D_1 are as before and the dependent variable is the recovered scale values (x_i). From Ming-Wei Wang (see the appendix), we know that b_1 and b_2 are correct estimates of the values sought only to a constant multiple (i.e., $b_1 = \theta \alpha$ and $b_2 = \theta \beta$). Once this factor has been determined, the true parameter estimates are known. θ is found through the use of the second stage of the algorithm, which involves a second regression analysis embedded in an iterative process. The goal of the iterative process is to find a θ that will transform the scale values x_i so that, when new distances are calculated, they will be correlated with the original distances with a regression coefficient of 1.0 and $r = 1.0$. This is done by working backward from the information gained from the regression in stage one. We calculate new scale values by:

$$x_1' = \theta \alpha \ln P_i + \theta \beta \ln D_i + a$$

and then new distances (S_{ij}'):

$$S_{ij}' = |(e^{x_i'}/e^{x_i} + e^{x_j}) - .05|$$

We regress these distances against the original distances as:

$$S'_{ij} = a + bs_{ij} + e_{ij}$$

We are searching for the situation that gives a $b = 1.0$ and an e_{ij} of a very small amount. Thus, we have a problem in which we search for a value for θ that will minimize e_{ij} above. Once θ is known, $\alpha = b_1\theta$ and $\beta = b_2\theta$, and we have the preference function values.

The steps in the algorithm of the second stage are: (1) Choose an arbitrary starting value for θ; choose a beginning step size by which θ will be varied, and call it Δ; and choose a tolerance level that will terminate the algorithm when the standard error of the regression coefficient is equal to or less than this value, and call it T. (2) Obtain new distances as above by using the current value of theta, and regress these new distances against the original distances. (3) Compare the standard error of the regression found in step 2 with the tolerance level T. If T is smaller, go on to step 4; otherwise, stop. (4) Compare the standard error of the regression with the error associated with the previous regression. If the most recent error is smaller than the previous error, increase theta by Δ and repeat step 2. If the most recent error is larger than the error from the previous regression, set Δ to $\frac{1}{2}\Delta$ and change θ to $\theta - \Delta$ and repeat step 2. The algorithm terminates with the correct parameters.

Testing the Procedure

The procedure described above was first tested on data with no error. This data was computed from the following preference function:

$$z_i = \frac{P_i^\alpha}{D_i^\beta}$$

and distances (S_{ij}):

$$S_{ij} = \left| \frac{z_i}{z_i + z_j} - 0.5 \right|$$

The combinations of distance and town size used to define forty-eight locational types are shown in Figure 1. The midpoint values for each locational type shown there were used to determine the z_i in the above equations.

Using these locational types and the preference function above, an errorless dissimilarities matrix was computed for input to the MDS routine TORSCA and a scale was computed [11]. This scale

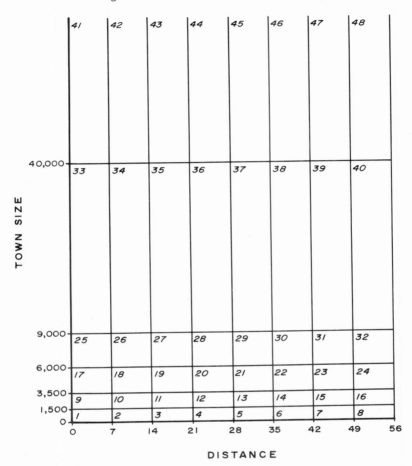

Figure 1. Stimulus groups used in the simulation experiments.

had a stress value of .092 associated with it. The Shepard diagram of these results is shown on Figure 2. The algorithm described previously was used and an α of 0.79 and a β of 1.99 were found. Compare this with the true values of $\alpha = 0.8$ and $\beta = 2.0$.

Encouraged by the quality of the results produced, we then tested the method on simulated spatial choice data. A data set of spatial choices was generated for 239 towns in southern Iowa. The choices were made from a regular grid of 658 points spaced 5 miles apart in the three southern tiers of counties in Iowa. Sixty-

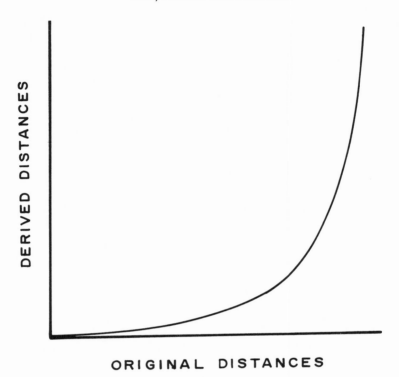

ORIGINAL DISTANCES

Figure 2. Shepard diagram from first experiment.

one choices were generated from each grid point moving from east to west until a grand sample of 40,000 choices was created.

Each choice was determined by first computing the probability of a town being chosen by an individual at one of the 658 sample grid points. Each grid point has, according to the preference function, a utility for each jth town of:

$$U_{ij} = P_j^\alpha / D_{ij}^\beta$$

and a probability of being assigned to the jth town of:

$$P_{ij} = U_{ij} / \sum_{j=1}^{n} U_{ij}$$

This decision model is the familiar choice axiom of Luce [4]. The spatial choice is made through a random selection mechanism with the likelihood of a choice being made directly proportional to the computed probability for that place. This random selection procedure is thus repeated sixty-one times from each grid point.

When a choice is made, others are rejected unless they are considered to be equivalent (i.e., a person is indifferent to two or more town size/distance combinations). Each choice can have allocated to it a person who selects that choice over all others. When such an assignment occurs, the information is recorded.

From the master sample of 40,000, smaller regular random samples were drawn, ranging in size from 200 to 15,000. Once a new sample was drawn, a new interstimulus distance matrix was computed according to the method described earlier and used as input to the MDS routine for the computation of the scale.

Results

A typical result estimated α to be 0.524 and β to be 1.60. (This should be compared with true values of 0.8 and 2.0, respectively.) The above estimates were computed by using the midpoint averages for each class of stimuli in the regression equation $(P + D)$ of the first phase of the algorithm. The use of the midpoints, rather than a more realistic figure, could certainly lead to specification error. Thus, three other averages were tried: (1) the actual averages of all stimuli in each locational type, (2) the average of all stimuli chosen in each locational type, and (3) the average of all stimuli rejected in each locational type. Results of using these four averages are shown in Figure 3.

Which of the four averages gives the best result? Midpoint averages clearly seem to give the worse results. Both α and β are better estimated by the other three methods. But which of these three is the best? There is no well-developed procedure for determining the best estimates. We determined which solution was best by using the estimated parameters to determine the probability that places would be chosen. For any individual, the sum of the differences between these probabilities and those computed from the original parameter values (divided by two) gives the expected systematic bias that would occur if the estimated parameter values were assumed to be correct. In the figure, these biases were computed for hypothetical pairs of parameter values and a continuous surface was interpolated to create percentage error contours. (See Figure 4.) When these are placed on the preceding graph, it can be seen that the chosen locational types give parameter estimates with the least amount of error. This error is a little over 2% and came from samples ranging in size from 2,000 to 15,000.

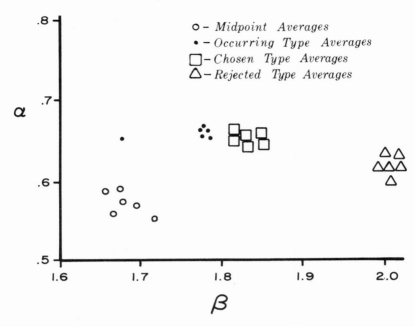

Figure 3. Relationship between recovered parameters and method of estimating individual stimulus values.

Samples this large are difficult and expensive to survey, and so it seemed important to determine how small a sample could be used without loss of accuracy. At sample sizes of 200 to 300, the parameter recovery deteriorated to the point where the method recovered parameters with the wrong sign. At sample sizes of 400, the method seemed to recover consistently parameters with an error of around 2%, and in some cases the error was slightly less.

CONCLUSION

These results show that it is possible to recover accurately the parameters of a decision model from hypothetical spatial choice data generated by applying the decision model to a sample of locations in a study area. These results can be compared with those of Sheppard [9], who showed that, when the attractiveness of destinations was variable, the spatial structure of an area resulted in biased estimates of the parameters of gravity models even when behavior conformed to the gravity hypothesis. In this study, however,

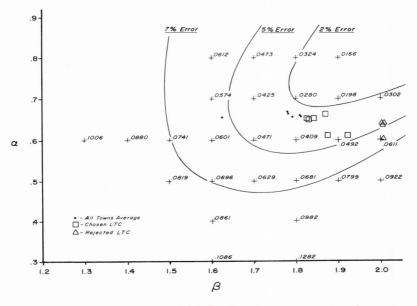

Figure 4. Error in predicting spatial choice given that true parameter values are $\alpha = .8$ and $\beta = 2.0$.

although no assumptions were made about the spatial structure of the study area (indeed, in full knowledge that an earlier study had shown that the occurrence of town-size attractiveness-distance combinations in Iowa are not independent of one another [7, pp. 399-400]), no bias was found in the estimates of the parameters of the decision model. This result confirms the claims for model robustness in the face of differences in the spatial structure of study areas that had been made for the revealed preference approach to the analysis of spatial choice.

APPENDIX

Recovering the Parameters of a Function That Generated Data for a Nonmetric MDS Scale

Let x_{ij} = number of times ith stimulus is chosen when both ith and jth stimuli are present:

This material was prepared by Dr. Ming-Mei Wang of the Rand Corporation, Santa Monica, Calif.

$$p_{ij} = \frac{x_{ij}}{x_{ij} + x_{ji}}$$

$$\delta_{ij} = |p_{ij} - 0.5| \qquad 0 < \delta_{ij} < 0.5$$

Then:

$$\frac{p_{ij}}{p_{ji}} = \frac{0.5 + \delta_{ij}}{0.5 - \delta_{ij}} \qquad\qquad p_{ij} > 0.5$$

$$\frac{p_{ij}}{p_{ji}} = \frac{0.5 - \delta_{ij}}{0.5 + \delta_{ij}} \qquad\qquad p_{ij} < 0.5$$

$$\ln p_{ij} - \ln p_{ji} = \ln \frac{0.5 + \delta_{ij}}{0.5 - \delta_{ij}} \qquad p_{ij} > 0.5$$

$$\ln p_{ij} - \ln p_{ji} = - \ln \frac{0.5 + \delta_{ij}}{0.5 - \delta_{ij}} \qquad p_{ij} < 0.5$$

$$|\ln p_{ij} - \ln p_{ji}| = \ln \frac{0.5 + \delta_{ij}}{0.5 - \delta_{ij}}$$

Note that $\ln p_{ij} - \ln p_{ji} = \ln z_i - \ln z_j$. Thus, define $d_{ij} = (a \ln z_i + b) - (a \ln z_j + b) = a(\ln z_i - \ln z_j)(a > 0, b$ constant). We have:

$$d_{ij} = a \ln \frac{0.5 + \delta_{ij}}{0.5 - \delta_{ij}} \qquad\qquad (1)$$

Since $0 \leqslant \delta_{ij} \leqslant 0.5$, equation 1 implies that d_{ij} is a monotonic increasing function of δ_{ij}, or we can write:

$$\delta_{ij} = 0.5 \frac{e^{d_{ij}/a} - 1}{e^{d_{ij}/a} + 1} \qquad d_{ij} \geqslant 0$$

$$= 0.5 - \frac{1}{e^{d_{ij}/a} + 1}$$

so that δ_{ij} is a monotonic increasing function of d_{ij}. This shows that what the nonmetric scaling program recovers is a linear function of $\ln z_i$ (if dimensionality is 1). That is, if we denote x_i to be the one-dimensional solution for inputs s_{ij}, we find that $x_i = \ln z_i + b (a \geqslant 0, a, b$ constants).

The perfect regression equation of x on z is:

$$x_i = a \ln z_i + b$$

$$= a[\alpha \ln p_i - \beta \ln D_i] + b$$

$$= a\alpha \ln p_i - a\beta \ln D_i + b$$

That is, if one does linear regressions of x_i on $\ln p_i$ and $\ln D_i$, the regression weights will be $a\alpha$, $a\beta$, respectively, for $\ln P_i$, $\ln D_i$, while the intercept will be b ($\neq 0$ in general). Thus, if one generates data with:

$$z_i = P_i^{\alpha}/D_i^{\beta}$$

where P_i = town size of the ith town and D_i = distance of the ith town ($\alpha = 0.8$, $\beta = 2$), then regression weights should be some multiples of 0.8 and 2, respectively, and intercept should be some value ($\neq 0$, in general).

REFERENCES

1. Hubbard, R., and A. F. Thompson, Jr., 1979. "Generalizations on the Analysis of Spatial Indifference Behavior," *Geographical Analysis*, 11, 196-201.
2. Johnston, R. J., 1976. "On Regression Coefficients in Comparative Studies of the Frictions of Distance," *Tijdschrift voor Economische en Sociale Geografie*, 67, 51-58.
3. Louviere, J. J., 1976. "Information-Processing Theory and Functional Form in Spatial Behavior." In R. G. Golledge and G. Rushton (eds.), *Spatial Choice and Spatial Behavior*, pp. 211-46. Columbus, Ohio: Ohio State University Press.
4. Luce, R. D., 1959. *Individual Choice Behavior: A Theoretical Analysis.* New York: John Wiley and Sons.
5. Olsson, G., 1970. "Explanation, Prediction and Meaning Variance: An Assessment of Distance Interaction Models," *Economic Geography*, 46, 223-33.
6. Rushton, G., 1969. "The Scaling of Locational Preferences." In K. R. Cox and R. G. Golledge (eds.), *Behavioral Problems in Geography: A Symposium*, pp. 197-227. Evanston, Ill.: Department of Geography, Northwestern University, Studies in Geography #17.
7. Rushton, G., 1969. "Analysis of Spatial Behavior by Revealed Space Preference," *Annals, Association of American Geographers*, 59, 391-400.
8. Sheppard, E. S., 1978. "Theoretical Underpinnings of the Gravity Hypothesis," *Geographical Analysis*, 10, 386-402.
9. Sheppard, E. S., 1979. "Gravity Parameter Estimation," *Geographical Analysis*, 11, 120-32.
10. Stetzer, F., and A. G. Phipps, 1977. "Spatial Choice Theory and Spatial Indifference: A Comment," *Geographical Analysis*, 9, 400-403.
11. Young, F. W., and W. S. Torgerson, 1967. "TORSCA, a FORTRAN IV Program for Shepard-Kruskal Multidimensional Scaling Analysis," *Behavioral Science*, 12, 498.

Chapter 2.3 A Revealed Preference Analysis of Intraurban Migration Choices

W. A. V. Clark

Although Rushton's [16] methodology for the analysis of spatial choices was published some time ago, there has been only limited application of the methodology in geography. Recently, there have been attempts to compare the spatial preference scaling model with a utility theory approach [19] and to extend it "by improving its mathematical and conceptual base" [7] and in particular to enlarge the number of locational types that might be considered in the model. Nevertheless, the empirical applications of the model have been only to consumer behavior [7, 16, 17], and its validity for investigating other choice situations has not been investigated.

The idea of incorporating preferences into explanations of migration between cities and regions and the attempt to analyze the preferences of migrants has been suggested by Gould [9] and explored by Demko [5], but there has been little progress in developing a methodology for this purpose. It seems reasonable, therefore, to extend the Rushton technique of revealed preferences to the analysis of the trade-offs that migrants make in moving within

This paper was prepared during the summer of 1976 and has been only slightly modified since that time. I would like to thank Gail Jensen of UCLA for the data preparation and Gerard Rushton and Jim Kohler for suggestions and a program for the data analysis. I would also like to thank the UCLA Office of Academic Computing for financial support.

a city. Although it is true at the present time that we have only limited knowledge "about how individuals in different real life situations perceive sets of spatial alternatives and how their perceptions might be related to their choice behavior" [3, p. 183], it does seem that the revealed preference methodology could be used as a technique for uncovering the relationships among alternatives and spatial choices. Further, until we have better information on how people structure choices, it is not unreasonable to use the simple assumption of trade-offs among stimuli.

It is true that the use of a trade-off function in which households evaluate attributes assumes that the attributes that influence the selections are relevant. Even though Burnett [3, p. 182] has forcefully argued that these need not be the real dimensions that influence choices, until we are sure that we know the true attributes and whether the attributes can be grounded in theory, it is not unreasonable to pursue the use of such attributes in a trade-off model. However, this apparent disagreement is at the heart of a division in the work of behavioral geographers, and it must be recognized explicitly.

Those investigations in geography that can be described loosely as behavioral investigations can be divided into studies that place the main emphasis on cognition, learning, and models of choice behavior and in which the data for empirical analyses are most often collected from laboratory or questionnaire responses to a variety of stimuli [3, 10, 11, 21]; and studies that are concerned with the analysis of actual behavior in space and the inferences about preferences that can be drawn therefrom [7, 16, 17]. In many instances, the distinction is also between a concern with an experimental analysis of individual responses to stimuli and the attempt to investigate stimuli in the actual setting in which consumers make choices. Each approach has its proponents, but there is an overlooked value in the revealed preference analysis of behavior— that is its potential for elucidating the links between structure and behavior. It has been argued that "the rules by which we make choices among alternatives in space are not modifiable by the form of the system" [17, p. 393] and, thus, that we can use the method to identify spatial behavior in a fundamental sense, and the spatial behavior in a system that is constrained by the particular system. This, of course, does not solve the problem—which are the relevant specific constraints of structure—but it may bring us a step closer to the elusive links between behavior and structure. This, after all, is one of the major goals of social science investigation. Now that there are increasing amounts of information, both about the detailed structural composition of urban areas and about the behavior of individuals (and also, in some cases, about the actual preferences and feelings of these individuals), it is useful to attack the problem of developing models of spatial choice and evaluation both from the experimental-laboratory analysis of preference behaviors and from the revealed preference evaluation of actual behaviors within a region or a city.

There is another reason for studying both preferences expressed through cognition and those derived from actual behaviors. Both Festinger [6] and Wicker [20] have argued that attitude (cognition) and behavior are not easily separated even though the assumption in many geographical studies is that attitude is a good predictor of behavior. Thus, experimental studies of a trade-off between distance and cost need to be paralleled by analyses of revealed preferences derived from actual consumer behavior. It is obvious and it is a truism that behavioral choice processes will not be easily unlocked, and no one approach is likely to provide all the answers. So far most of the work on preference analysis has been of the experimental trade-off variety, and the use of the revealed preference methodology has had a much more limited application. However, if it is to be validated as a general methodology, it must be applied in contexts other than that of consumer behavior. The present application is a small step in the validation procedure in which data on intraurban migration choices are used as input to a revealed preference analysis.

LOCATIONAL CHOICE

Rushton [16, p. 198] has noted that "spatial behavior implies a search among alternatives." It is just such a search procedure that is undertaken by the intraurban migrant who decides to change residences [2]. Thus, we can imagine a process of spatial choice in which a potential mover "compares each alternative with every other one and selects that which he expects will give him the greatest satisfaction" [16, p. 198]. In his research Rushton was able to construct an ordinal preference function that indicated "which of several alternatives would give the greatest satisfaction" [16, p. 198]. With the assumptions "that the decisions of different people are generated from similar preference functions" and that it is possible to uncover the subjective preference function from the analysis of a large number of different individual decisions [16, p. 199], it is possible to establish a ranking of the choice alternatives.

The subjective preference function ranks an individual's preference for a set of objects [16, p. 199]. In his analysis of consumer choices, Rushton focused on a variety of town sizes (as surrogates for the functional offerings of towns) and their distances from each consumer. Thus, "the relevant properties of towns are . . . summarized . . . in town population and the spatial relationship of interest is the separation between the individual and the town" [16, p. 200]. Thus, an individual consumer chooses a locational type that is defined as a combination of size and distance from the location of his or her residence. The important point of this format is that it takes the unique spatial context and converts it to a more general locational type.

When a household is concerned about choosing a new location, the consumer ordinarily evaluates a small or large number of houses, selects one of these, and rejects the others [2, 13]. There are, of course, many relevant properties that can be evaluated by movers, and the number of opportunities within the city is very large (all houses and apartment units). However, there are several models that suggest the ways in which movers make the trade-offs among the attributes of housing and location, in their relocation decision making. One basic and well-developed model describes residential choices in terms of a trade-off between transportation costs (usually to the central business district) and the price of housing. Alonso's [1] model is best described as the comparison of three goods—amount of space (which is related to the price of housing), transportation costs, and expenditures on a composite good (all other household expenditures). This model is quite simplistic and ignores a variety of elements of urban structure, in particular the fact that rent and transportation gradients are not constant. On the other hand, they are suggestive of the relevant properties in housing choices [14]. Although the Alonso [1], Muth [12], and Wingo [21] models emphasize the importance of accessability in determining location, several other writers have argued for the importance of amenities in residential choice [15, 18]. A recent paper [14] has argued that *"both* accessibility and amenities are important to the residential location decision," and this notion accords well with intuitive notions of the trade-offs in which intraurban movers engage. As Pleeter [14, p. 374] noted, "where access is equal variations in amenities will be the primary force in locational choice; where amenities are equal it will be accessibility which determines location."

Locational types are not readily defined for individual houses, and in this study the census tract is used as a sufficiently small unit to define a locational type. The first component of the locational type is distance from the central business district. This component represents the accessibility component of the Alonso-type models. The second component is amenity and is taken as the population density of the tract. Density is used as a surrogate for both housing space and size and for the amount of open space in the neighborhood. High-density tracts are assumed on the whole to have crowded dwelling units, small dwellings, and small amounts of open space; and areas with low densities are assumed to be the opposite. Thus, households are trading off accessibility against amenity in their moves from one location to another. (See Figure 1.) It is true that the central business district may not be the node to which the movers are responding. A modification to be discussed later will allow for accessibility to work place to be an important constraint on the relocation choice, but it is worth noting that in 1960 Milwaukee was a particularly nodal city, with many of its city services, opportunities, and jobs located in the central business district.

Unlike the case of consumer choices, the number of opportunities for a mover is not simple. In the Rushton study, consumers evaluated towns of specified sizes at a variety of distances within a 28-mile radius from their location. One town was selected for the expenditures on grocery purchases, and others were rejected. Not all opportunities were available to all consumers simply because of the spatial arrangement of towns in a region. In the case of intra-urban movers, two different constraints were investigated. The first test of revealed preferences is developed by taking each locational type that is within 6 miles of the home location of a mover prior to relocation. Six miles includes 92% of all moves made in Milwaukee in 1960-1962. (See Table 1.) A second constraint was to use the journey to work as a limit to the choices. All locational types that were within the radius of the journey to work (based on the work place) were considered as potential choices for the movers. A redefinition of locational types in terms of density and distance from the work place is also possible. Thus, if the distance to work were 1.8 miles, all tracts (and their densities) within that radius were considered as potential sites for relocation. The locational

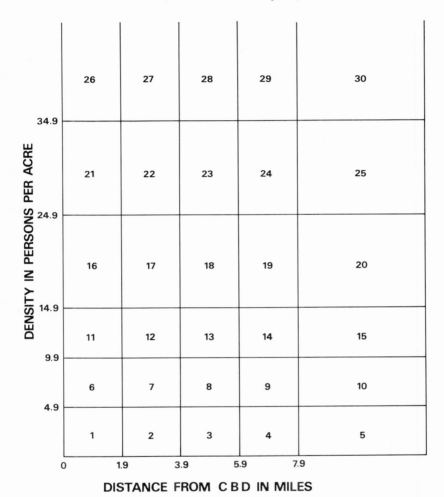

Figure 1. Definition of locational types.

types for that mover would be densities within ranges of 1 and 2 miles. The present data are not sufficiently detailed for such an analysis. Partial presentations of the choice data are given for each constraint in Tables 2 and 3.

The data used to generate the trade-offs between accessibility and amenity are drawn from a large random sample of households in southeastern Wisconsin for the period 1950 to 1963. Of some

TABLE 1
Percentage of Households Moving Specified Distances

Distance in Miles	Number of Movers	Percentage
0-0.5	23	5.11
0.5-1.0	69	15.33
1.0-1.5	59	13.11
1.5-2.0	56	12.44
2.0-2.5	29	6.44
2.5-3.0	41	9.11
3.0-3.5	26	5.77
3.5-4.0	27	6.00
4.0-4.5	23	5.11
4.5-5.0	19	4.22
5.0-5.5	31	6.88
5.5-6.0	9	2.00
6.0-6.5	11	2.44
6.5-7.0	3	.66
7.0-7.5	2	.44
7.5-8.0	5	1.11
8.0-8.5	7	1.55
8.5-9.0	4	.88
9.0-9.5	1	.22
9.5-10.0	4	.88
10.0 +	1	.22

16,000 households, 9000 moved sometime during this period; from that file 450 households that moved once and specifically during the period 1960 to 1962 were chosen for analysis with the revealed preference methodology. The second data set was the distances of all tracts from the CBD and the population density per acre in 1960. The third data set was the place of work for each household in 1960.

PAIRED COMPARISONS AND PERCEIVED SIMILARITY OF LOCATIONAL TYPES

Rushton has already given a clear exposition of the method of paired comparisons and the way in which it can be used to derive the input of any scaling analysis, and it is not necessary to repeat that presentation here. The basic data, which are the actual choices made by the movers and their rejected alternatives (for each constraint), are shown in Tables 2 and 3. From this data a matrix of

Mover ID	1	2	3	4	5	6	7	8	9	10	11	12	13	14	15	16	17	18	19	20	21	22	23	24	25	26	27	28	29	30
1	*	*	*	*	*	*	*	*	*	*	*	*	*	1		*	*	*	*	*	*	*	*	*	*	*	*			
2		*	*	1	*	*	*	*	*	*	*	*	*		*	*	*	*	*	*	*	*	*	*						
3			*	*	*	*	*	*	*	*	*	*	*	1		*	*	*	*	*	*	1	*		*	*	*			
4	*	*	*	*	*	*	*	*	*	*	*	*	*	*		*	*	*	*	*	*	*	*		*	*	*			
5	*	*	*	*	*	*	*	*	*	*	*	1	*	*		*	*	*	*	*	*	*	*		*	*	*			
6	*	*	*	*	*	1	*	*	*	*	*	*	*	*		*	*	*	*	*	1	*	*		1	*	*			
7	*	*	*	*	*	*	*	*	*	*	*	*	*	*		*	*	*	*	*	*	*	*		*	*	*			
8	*	*	*	*	*	*	*	*	*	*	*	*	*	1		*	*	*	*	*	*	*	*		*	*	*			
9	*	*	*	*	*	*	*	*	*	*	*	*	*	1		*	*	*	*	*	*	*	*		*	*	*			
10	*	*	*	*	*	*	*	*	*	*	*	*	*	*		*	*	*	*	*	*	*	*		*	*	*			
11	*	*	*	*	*	*	*	*	*	*	*	*	*	*		*	*	*	1	*	*	1	*		*	*	*			
12	*	*	*	1	*	*	*	*	*	*	*	*	*	*		*	*	1	*	*	*	*	*		*	*	*			
13	*	*	*	1	*	*	*	*	*	*	*	*	*	*		*	1	*	*	*	*	*	*		*	*	*	1		
14	*	*	*	1	*	*	*	*	*	*	*	*	*	*		*	*	*	*	*	*	*	*		*	*	*			
15		*	*	*	*	*	*	*	*	*	*	*	*	*	*	*	*	*	*	*	*	*	*		*	*	*			
16			*	*	*	*	1	*	*	*	*	*	*	*		*	1	*	*	*	*	*	*		*	*	*			
17			*	*	*	*	1	*	*	*	*	*	*	*		*	*	*	*	*	*	*	*		*	*	*			
18			*	*	*	*	*	*	*	*	*	*	*	*		*	1	1	*	*	*	*	*		*	*	1			
19			*	*	*	*	*	*	*	*	*	*	*	*		1	*	*	*	*	*	*	*		*	*	*			
20			*	*	*	*	*	*	*	*	*	*	*	1		1	1	*	*	*	*	*	*		*	*	*			
21			*	*	*	*	*	*	*	*	1	1	*	*		1	1	*	*	*	*	*	*		*	*	1			
22			*	*	*	*	*	*	*	*	1	*	*	*	*	*	*	*	1	*	*	*	*	1	*	*	*			
23					*										*				1											
24					*	1									*									1						
25			*	*	*	*	*	*	*	*	*	*	*	1		*	*	1	*	*	*	*	*		*	*	*			
26			*	*	*	*	*	*	*	*	*	*	*	*		*	*	1	*	*	*	*	*		*	*	*			
27			*	*	*	*	*	1	*	*	*	*	*	*		*	*	*	*	*	*	1	*		*	*	*			
28			*	*	*	*	*	*	*	*	*	*	*	*		1	*	*	*	*	*	*	*	*	*	*	*			
29			*	*	*	*	*	*	*	*	*	*	*	*		*	*	*	*	*	*	*	*	*	*	*	*			
30			*	1	*	*	*	*	*	*	*	*	*	*		*	*	*	*	*	*	*	*	*	*	*	*			

NOTE: Locational type chosen is indicated by a 1. Locational type available is indicated by an *. (A type was considered available when it was within 6 miles of the original location of the mover.) A blank indicates an unavailable locational type.

TABLE 3
Revealed Space Preference, Raw Data Matrix by Movers II

Mover ID	1	2	3	4	5	6	7	8	9	10	11	12	13	14	15	16	17	18	19	20	21	22	23	24	25	26	27	28	29	30
1				*									1																	
2																					1									
3						*						1																		
4						*										*				1		*				*	*			
5	*	*	*	*		*	*	*	*	*	*	*	1			*	*	*	*		*	*	*	*		*	*			
6													1																	
7	*	*	*		*	*	*	*	*	*	*	*	*	*	*	*	*	*	1		*	*	*			*	*			
8	*	*				*										*					*	1				*				
9	*	*		*	1	*	*	*			*	*	*			*	*	*			*	*	*	*		*	*			
10					1																									
11				*	1		*				*		*				*	*					*	*						
12				*		1	*				*							*						*						
13	*	*				*	1									*					*					*	*			
14							*				*		*				1	*												
15	*	*				*					*		*			*	1				*					*				
16	*	*	*			*	*	*			*	*	*			*	*	*	*		*	*	*	*	1	*	*			
17				1											*															
18	*	*				*	*				*	*	*	1		*	*	*			*	*	*			*	*			
19																		1												
20						1					*		*				*	*												
21		*	*		*				*	*		*			*				*		1									
22																	1													
23				1																										
24	*	*	*	*	1	*	*	*			*	*	*	*		*	*	*	*		*	*	*	*		*	*			
25	*	*	*			*	*				*	*	*			*	*	1			*	*	*			*	*			
26	*	*				*	*				*		*	1		*	*	*			*	*	*			*	*			
27	*	*	*		*	*	*		*	*	*	*	*	1	*	*	*		*		*	*	*			*	*			
28						*						1																		
29	*	*				*	*				*	*		*		*	1		*		*	*	*			*	*			
30	*	*				*	1									*					*	*				*	*			

NOTE: Locational type chosen is indicated by a 1. Locational type available is shown by an *. (A type was considered available when it was within the circle defined by the journey to work. The circle was centered over the work place.) A blank indicates an unavailable locational type.

similarities between any two locational types can be developed. This "measure of similarity . . . is the degree to which one is preferred by persons who can choose both types . . . and . . . is the proportion of times one type is chosen over the other when both are present" [16, p. 205]. Tables 4 and 5 show the probability, P_{jpi}, that the jth locational type is preferred to the ith locational type. The degree to which any two locational types are similar is given by the value 0.5 in Tables 4 and 5. These tables can be translated into perceived dissimilarity measures by taking the absolute value of P_{jpi} − 0.5.

SCALING THE PROXIMITY MATRICES

The heart of the revealed preference methodology lies in the ability to construct a scale, from these proximity values, along which all the locational types will be arranged. As Rushton [16, p. 209] notes, if a scale that orders the locational types can be constructed, then this "would demonstrate that the spatial choice data can be regarded as having been generated from a preference structure, since preference structures are an ordering of all conceivable alternatives." There are several algorithms for deriving such a scale. Torgerson's multidimensional scaling program was applied to the data in Tables 6 and 7, and the computed coordinates are given in Tables 8 and 9. The stress for the scale from the first proximity matrix is 0.176 and for the second somewhat higher at 0.437. In both cases, Torgerson's index value was high—0.99 and 0.95, respectively. (A value of 0.9 is good and 0.5, poor.) Thus, a preliminary review of the results suggests that intraurban migrant choices can be ordered and a preference function derived from the scale. It is in attempting to elucidate the preference structure that some difficulties arise.

In Figures 2 and 3 an attempt is made to construct an isopleth for the coordinate data. Although it is possible to interpolate lines that are of equal scale value, it is apparent that it is not as straightforward as in either Rushton [16] or Girt [7]. It is possible that this is a function of the scaling and data input or that it is revealing of peculiarities of migrant preferences. Some observations will be made on each of these possibilities. The shape of the indifference map for the first set of proximity coordinates is expected. That is, migrants are seemingly indifferent when choosing between low densities at long distances from the CBD and high densities at closer distances. In essence, this is the argument of the Alonso [1],

TABLE 4

Probability that Column Locational Type Is Preferred to Row Type, from Raw Data Matrix I

						Locational Type								
	1	2	3	4	5	6	7	8	9	10	11	12	13	14
1	0.0	1.00	1.00	1.00	1.00	1.00	1.00	1.00	1.00	1.00	1.00	1.00	1.00	1.00
2	0.0	0.0	0.53	0.77	0.81	0.13	0.53	0.72	0.63	0.40	0.22	0.46	0.83	0.74
3	0.0	0.47	0.0	0.78	0.81	0.10	0.40	0.71	0.68	0.67	0.25	0.44	0.82	0.83
4	0.0	0.23	0.22	0.0	0.60	0.04	0.14	0.43	0.38	0.36	0.10	0.19	0.57	0.49
5	0.0	0.19	0.19	0.40	0.0	0.0	0.06	0.33	0.32	0.27	0.07	0.15	0.45	0.45
6	0.0	0.88	0.90	0.96	1.00	0.0	0.89	0.95	0.94	1.00	0.75	0.88	0.97	0.96
7	0.0	0.47	0.60	0.86	0.94	0.11	0.0	0.71	0.78	0.90	0.27	0.44	0.82	0.88
8	0.0	0.28	0.29	0.57	0.67	0.05	0.29	0.0	0.49	0.56	0.13	0.24	0.66	0.62
9	0.0	0.38	0.32	0.63	0.68	0.06	0.22	0.51	0.0	0.43	0.06	0.29	0.68	0.62
10	0.0	0.60	0.33	0.64	0.73	0.0	0.10	0.44	0.57	0.0	0.11	0.23	0.60	0.71
11	0.0	0.78	0.75	0.90	0.93	0.25	0.73	0.88	0.94	0.89	0.0	0.70	0.95	0.89
12	0.0	0.54	0.56	0.81	0.85	0.13	0.56	0.76	0.71	0.77	0.30	0.0	0.85	0.83
13	0.0	0.17	0.18	0.43	0.55	0.03	0.18	0.34	0.32	0.40	0.05	0.15	0.0	0.45
14	0.0	0.26	0.17	0.51	0.55	0.04	0.12	0.38	0.38	0.29	0.11	0.17	0.55	0.0
15	−1.00	1.00	1.00	1.00	1.00	−1.00	1.00	1.00	1.00	1.00	1.00	1.00	1.00	1.00
16	0.0	0.47	0.53	0.77	0.90	0.11	0.47	0.70	0.75	0.80	0.27	0.40	0.80	0.80
17	0.0	0.16	0.19	0.44	0.61	0.02	0.18	0.34	0.33	0.63	0.07	0.15	0.52	0.48
18	0.0	0.18	0.17	0.45	0.45	0.03	0.17	0.34	0.34	0.30	0.05	0.15	0.51	0.47
19	0.0	0.32	0.38	0.70	0.76	0.06	0.17	0.53	0.52	0.62	0.12	0.29	0.69	0.67
20	−1.00	−1.00	−1.00	−1.00	−1.00	−1.00	−1.00	1.00	1.00	1.00	1.00	1.00	−1.00	−1.00
21	0.0	0.24	0.30	0.57	0.81	0.04	0.27	0.45	0.44	0.80	0.08	0.24	0.65	0.61
22	0.0	0.15	0.19	0.42	0.57	0.02	0.19	0.36	0.31	0.39	0.07	0.15	0.51	0.46
23	0.0	0.54	0.54	0.81	0.81	0.13	0.53	0.71	0.74	0.62	0.22	0.42	0.84	0.81
24	0.0	0.80	0.89	0.90	0.93	0.0	0.88	0.86	0.82	0.80	0.67	0.75	0.97	0.95
25	−1.00	−1.00	−1.00	−1.00	−1.00	−1.00	−1.00	1.00	1.00	1.00	1.00	−1.00	1.00	−1.00
26	0.0	0.28	0.33	0.64	0.84	0.05	0.30	0.51	0.50	0.67	0.10	0.24	0.68	0.64
27	0.0	0.19	0.22	0.48	0.71	0.03	0.20	0.38	0.34	0.54	0.03	0.15	0.54	0.48
28	−1.00	−1.00	−1.00	−1.00	−1.00	−1.00	−1.00	−1.00	−1.00	−1.00	−1.00	−1.00	−1.00	−1.00
29	−1.00	−1.00	−1.00	−1.00	−1.00	−1.00	−1.00	−1.00	−1.00	−1.00	−1.00	−1.00	−1.00	−1.00
30	−1.00	−1.00	−1.00	−1.00	−1.00	−1.00	−1.00	−1.00	−1.00	−1.00	−1.00	−1.00	−1.00	−1.00

TABLE 4 (Continued)

						Locational Type									
6	17	18	19	20	21	22	23	24	25	26	27	28	29	30	
00	1.00	1.00	1.00	−1.00	1.00	1.00	1.00	1.00	−1.00	1.00	1.00	−1.00	−1.00	−1.00	1
53	0.84	0.82	0.68	−1.00	0.76	0.85	0.46	0.20	−1.00	0.72	0.81	−1.00	−1.00	−1.00	2
47	0.81	0.83	0.62	−1.00	0.70	0.81	0.46	0.11	−1.00	0.67	0.77	−1.00	−1.00	−1.00	3
23	0.56	0.55	0.30	−1.00	0.43	0.58	0.19	0.10	−1.00	0.36	0.52	−1.00	−1.00	−1.00	4
10	0.39	0.55	0.24	−1.00	0.19	0.43	0.19	0.07	−1.00	0.16	0.29	−1.00	−1.00	−1.00	5
89	0.97	0.97	0.94	−1.00	0.96	0.97	0.88	1.00	−1.00	0.95	0.97	−1.00	−1.00	−1.00	6
53	0.82	0.83	0.83	−1.00	0.73	0.81	0.47	0.13	−1.00	0.70	0.80	−1.00	−1.00	−1.00	7
30	0.66	0.66	0.47	0.0	0.55	0.64	0.29	0.14	0.0	0.49	0.62	−1.00	−1.00	−1.00	8
25	0.67	0.66	0.48	0.0	0.56	0.69	0.26	0.18	0.0	0.50	0.66	−1.00	−1.00	−1.00	9
20	0.38	0.70	0.38	0.0	0.20	0.61	0.38	0.20	0.0	0.33	0.46	−1.00	−1.00	−1.00	10
73	0.93	0.95	0.88	0.0	0.92	0.93	0.78	0.33	0.0	0.90	0.97	−1.00	−1.00	−1.00	11
60	0.85	0.85	0.71	0.0	0.76	0.85	0.58	0.25	−1.00	0.76	0.85	−1.00	−1.00	−1.00	12
20	0.48	0.49	0.31	−1.00	0.35	0.49	0.16	0.03	0.0	0.32	0.46	−1.00	−1.00	−1.00	13
20	0.52	0.53	0.32	−1.00	0.39	0.54	0.19	0.05	−1.00	0.36	0.52	−1.00	−1.00	−1.00	14
00	1.00	1.00	−1.00	−1.00	1.00	1.00	−1.00	1.00	−1.00	−1.00	1.00	−1.00	−1.00	−1.00	15
00	0.80	0.80	0.72	0.0	0.71	0.81	0.47	0.14	0.0	0.68	0.80	−1.00	−1.00	−1.00	16
20	0.0	0.50	0.31	−1.00	0.36	0.51	0.17	0.03	0.0	0.33	0.47	−1.00	−1.00	−1.00	17
20	0.50	0.0	0.31	0.0	0.37	0.49	0.14	0.03	0.0	0.34	0.48	−1.00	−1.00	−1.00	18
28	0.69	0.69	0.0	−1.00	0.53	0.72	0.29	0.12	−1.00	0.50	0.69	−1.00	−1.00	−1.00	19
00	−1.00	1.00	−1.00	0.0	1.00	−1.00	−1.00	−1.00	−1.00	1.00	−1.00	−1.00	−1.00	−1.00	20
29	0.64	0.63	0.47	0.0	0.0	0.64	0.24	0.05	0.0	0.46	0.60	−1.00	−1.00	−1.00	21
19	0.49	0.51	0.28	−1.00	0.36	0.0	0.17	0.04	−1.00	0.34	0.47	−1.00	−1.00	−1.00	22
53	0.83	0.86	0.71	−1.00	0.76	0.83	0.0	0.14	−1.00	0.73	0.82	−1.00	−1.00	−1.00	23
86	0.97	0.97	0.88	−1.00	0.95	0.96	0.86	0.0	−1.00	0.94	0.97	−1.00	−1.00	−1.00	24
00	1.00	1.00	−1.00	−1.00	1.00	−1.00	−1.00	−1.00	0.0	1.00	−1.00	−1.00	−1.00	−1.00	25
32	0.67	0.66	0.50	0.0	0.54	0.66	0.27	0.06	0.0	0.0	0.65	−1.00	−1.00	−1.00	26
20	0.53	0.52	0.31	−1.00	0.40	0.53	0.17	0.03	−1.00	0.35	0.0	−1.00	−1.00	−1.00	27
00	−1.00	−1.00	−1.00	−1.00	−1.00	−1.00	−1.00	−1.00	−1.00	−1.00	−1.00	0.0	−1.00	−1.00	28
00	−1.00	−1.00	−1.00	−1.00	−1.00	−1.00	−1.00	−1.00	−1.00	−1.00	−1.00	−1.00	0.0	−1.00	29
00	−1.00	−1.00	−1.00	−1.00	−1.00	−1.00	−1.00	−1.00	−1.00	−1.00	−1.00	−1.00	−1.00	0.00	30

TABLE 5
Probability That Column Locational Type is Preferred to Row Type, from Raw Data Matrix II

	Locational Type													
	1	2	3	4	5	6	7	8	9	10	11	12	13	14
1	0.0	1.00	1.00	1.00	1.00	−1.00	1.00	1.00	1.00	1.00	−1.00	−1.00	1.00	1.00
2	0.0	0.0	1.00	0.81	1.00	0.0	0.43	0.75	1.00	1.00	0.0	−1.00	0.80	1.00
3	0.0	0.0	0.0	1.00	1.00	0.0	0.0	0.50	0.75	1.00	0.0	0.0	0.86	0.90
4	0.0	0.19	0.0	0.0	1.00	0.0	0.07	0.13	0.25	1.00	0.0	0.0	0.18	0.24
5	0.0	0.0	0.0	0.0	0.0	0.0	0.0	0.0	0.0	0.30	0.0	0.0	0.05	0.05
6	−1.00	1.00	1.00	1.00	1.00	0.0	1.00	1.00	1.00	1.00	−1.00	−1.00	1.00	1.00
7	0.0	0.57	1.00	0.93	1.00	0.0	0.0	0.89	1.00	1.00	0.0	1.00	0.94	0.94
8	0.0	0.25	0.50	0.87	1.00	0.0	0.11	0.0	0.80	1.00	0.0	0.0	0.50	0.88
9	0.0	0.0	0.25	0.75	1.00	0.0	0.0	0.20	0.0	1.00	0.0	0.0	0.25	0.57
10	0.0	0.0	0.0	0.0	0.70	0.0	0.0	0.0	0.0	0.0	0.0	0.0	0.0	0.20
11	−1.00	1.00	1.00	1.00	1.00	−1.00	1.00	1.00	1.00	1.00	0.0	−1.00	1.00	1.00
12	−1.00	−1.00	1.00	1.00	1.00	−1.00	0.0	1.00	1.00	1.00	−1.00	0.0	1.00	1.00
13	0.0	0.20	0.14	0.82	0.95	0.0	0.06	0.50	0.75	1.00	0.0	0.0	0.0	0.74
14	0.0	0.0	0.10	0.76	0.95	0.0	0.06	0.13	0.43	0.80	0.0	0.0	0.26	0.0
15	1.00	−1.00	−1.00	−1.00	1.00	−1.00	−1.00	−1.00	−1.00	1.00	−1.00	−1.00	−1.00	1.00
16	0.0	0.75	1.00	0.94	1.00	0.0	0.75	1.00	1.00	1.00	0.0	−1.00	1.00	0.94
17	0.0	0.27	0.33	0.88	0.95	0.0	0.17	0.67	0.86	0.80	0.0	0.0	0.79	0.78
18	0.0	0.18	0.25	0.88	1.00	0.0	0.13	0.43	1.00	1.00	0.0	0.0	0.50	0.73
19	0.0	0.0	0.22	0.82	0.90	0.0	0.0	0.11	0.43	0.67	0.0	0.0	0.18	0.60
20	−1.00	−1.00	−1.00	−1.00	1.00	−1.00	−1.00	−1.00	−1.00	−1.00	−1.00	−1.00	−1.00	−1.00
21	0.0	−1.00	1.00	0.88	1.00	0.0	0.75	0.77	1.00	1.00	0.0	−1.00	0.85	1.00
22	0.0	0.0	0.57	0.81	0.88	0.0	0.20	0.67	0.67	0.57	0.0	0.0	0.78	0.81
23	0.0	0.44	0.60	1.00	0.96	0.0	0.33	0.83	0.89	0.83	0.0	0.0	0.76	0.93
24	−1.00	1.00	−1.00	1.00	1.00	−1.00	1.00	1.00	1.00	1.00	−1.00	−1.00	1.00	1.00
25	0.0	0.0	0.0	−1.00	1.00	0.0	0.0	0.0	−1.00	−1.00	0.0	0.0	0.0	−1.00
26	0.0	1.00	1.00	1.00	1.00	0.0	0.75	1.00	1.00	1.00	0.0	−1.00	1.00	0.94
27	0.0	0.0	0.71	1.00	1.00	0.0	0.27	0.91	1.00	1.00	0.0	0.0	0.94	0.76
28	−1.00	−1.00	−1.00	−1.00	1.00	−1.00	−1.00	−1.00	−1.00	−1.00	−1.00	−1.00	−1.00	−1.00
29	−1.00	−1.00	−1.00	−1.00	−1.00	−1.00	−1.00	−1.00	−1.00	−1.00	−1.00	−1.00	−1.00	−1.00
30	−1.00	−1.00	−1.00	−1.00	1.00	−1.00	−1.00	−1.00	−1.00	−1.00	−1.00	−1.00	−1.00	−1.00

TABLE 5 (Continued)

17	18	19	20	21	22	23	24	25	26	27	28	29	30	
						Locational Type								
1.00	1.00	1.00	-1.00	1.00	1.00	1.00	-1.00	1.00	1.00	1.00	-1.00	-1.00	-1.00	1
0.73	0.82	1.00	-1.00	-1.00	1.00	0.56	0.0	1.00	0.0	1.00	-1.00	-1.00	-1.00	2
0.67	0.75	0.78	-1.00	0.0	0.43	0.40	-1.00	1.00	0.0	0.29	-1.00	-1.00	-1.00	3
0.13	0.13	0.18	-1.00	0.13	0.19	0.0	0.0	-1.00	0.0	0.0	-1.00	-1.00	-1.00	4
0.05	0.0	0.10	0.0	0.0	0.13	0.04	0.0	0.0	0.0	0.0	0.0	-1.00	0.0	5
1.00	1.00	1.00	-1.00	1.00	1.00	1.00	-1.00	1.00	1.00	1.00	-1.00	-1.00	-1.00	6
0.83	0.87	1.00	-1.00	0.25	0.80	0.67	0.0	1.00	0.25	0.73	-1.00	-1.00	-1.00	7
0.33	0.57	0.89	-1.00	0.23	0.33	0.17	0.0	1.00	0.0	0.09	-1.00	-1.00	-1.00	8
0.14	0.0	0.57	-1.00	0.0	0.33	0.11	0.0	-1.00	0.0	0.0	-1.00	-1.00	-1.00	9
0.20	0.0	0.33	-1.00	0.0	0.43	0.17	0.0	-1.00	0.0	0.0	-1.00	-1.00	-1.00	10
1.00	1.00	1.00	-1.00	1.00	1.00	1.00	-1.00	1.00	1.00	1.00	-1.00	-1.00	-1.00	11
1.00	1.00	1.00	-1.00	-1.00	1.00	1.00	-1.00	1.00	-1.00	1.00	-1.00	-1.00	-1.00	12
0.21	0.50	0.82	-1.00	0.15	0.22	0.24	0.0	1.00	0.0	0.06	-1.00	-1.00	-1.00	13
0.22	0.27	0.40	-1.00	0.0	0.19	0.07	0.0	-1.00	0.06	0.24	-1.00	-1.00	-1.00	14
1.00	-1.00	1.00	-1.00	-1.00	1.00	1.00	-1.00	-1.00	-1.00	-1.00	-1.00	-1.00	-1.00	15
1.00	1.00	1.00	-1.00	0.75	1.00	0.83	0.0	1.00	0.25	0.91	-1.00	-1.00	-1.00	16
0.0	0.85	0.82	-1.00	0.21	0.46	0.31	0.0	1.00	0.0	0.17	-1.00	-1.00	-1.00	17
0.15	0.0	0.90	-1.00	0.18	0.25	0.11	0.0	1.00	0.0	0.07	-1.00	-1.00	-1.00	18
0.18	0.10	0.0	-1.00	0.0	0.36	0.10	0.0	1.00	0.0	0.0	-1.00	-1.00	-1.00	19
-1.00	-1.00	-1.00	0.0	-1.00	-1.00	-1.00	-1.00	-1.00	-1.00	-1.00	-1.00	-1.00	-1.00	20
0.79	0.82	1.00	-1.00	0.0	1.00	0.63	0.0	1.00	0.0	0.89	-1.00	-1.00	-1.00	21
0.54	0.75	0.64	-1.00	0.0	0.0	0.30	0.0	1.00	0.0	0.42	-1.00	-1.00	-1.00	22
0.69	0.89	0.90	-1.00	0.38	0.70	0.0	-1.00	1.00	0.17	0.55	-1.00	-1.00	-1.00	23
1.00	1.00	1.00	-1.00	1.00	1.00	-1.00	0.0	1.00	-1.00	-1.00	-1.00	-1.00	-1.00	24
0.0	0.0	0.0	-1.00	0.0	0.0	0.0	0.0	0.0	0.0	0.0	-1.00	-1.00	-1.00	25
1.00	1.00	1.00	-1.00	1.00	1.00	0.83	-1.00	1.00	0.0	1.00	-1.00	-1.00	-1.00	26
0.83	0.93	1.00	-1.00	0.11	0.58	0.45	-1.00	1.00	0.0	0.0	-1.00	-1.00	-1.00	27
-1.00	-1.00	-1.00	-1.00	-1.00	-1.00	-1.00	-1.00	-1.00	-1.00	-1.00	0.00	-1.00	-1.00	28
-1.00	-1.00	-1.00	-1.00	-1.00	-1.00	-1.00	-1.00	-1.00	-1.00	-1.00	-1.00	0.0	-1.00	29
-1.00	-1.00	-1.00	-1.00	-1.00	-1.00	-1.00	-1.00	-1.00	-1.00	-1.00	-1.00	-1.00	0.0	30

TABLE 6
Perceived Similarity between Locational Types—Proximity Matrix, from Raw Data Matrix I
(Absolute Difference of Probabilities from 0.5 with −1 for Missing Data)

						Locational Type								
	1	2	3	4	5	6	7	8	9	10	11	12	13	14
1	0.50	0.50	0.50	0.50	0.50	0.50	0.50	0.50	0.50	0.50	0.50	0.50	0.50	0.50 −
2	0.50	0.50	0.03	0.27	0.31	0.38	0.03	0.22	0.13	0.10	0.28	0.04	0.33	0.24 −
3	0.50	0.03	0.50	0.28	0.31	0.40	0.10	0.21	0.18	0.17	0.25	0.06	0.32	0.33
4	0.50	0.27	0.28	0.50	0.10	0.46	0.36	0.07	0.13	0.14	0.40	0.31	0.07	0.01
5	0.50	0.31	0.31	0.10	0.50	0.50	0.44	0.17	0.18	0.23	0.43	0.35	0.05	0.05
6	0.50	0.38	0.40	0.46	0.50	0.50	0.39	0.45	0.44	0.50	0.25	0.38	0.47	0.46 −
7	0.50	0.03	0.10	0.36	0.44	0.39	0.50	0.21	0.28	0.40	0.23	0.06	0.32	0.38
8	0.50	0.22	0.21	0.07	0.17	0.45	0.21	0.50	0.01	0.06	0.38	0.26	0.16	0.12
9	0.50	0.13	0.18	0.13	0.18	0.44	0.28	0.01	0.50	0.07	0.44	0.21	0.18	0.12
10	0.50	0.10	0.17	0.14	0.23	0.50	0.40	0.06	0.07	0.50	0.39	0.27	0.10	0.21
11	0.50	0.28	0.25	0.40	0.43	0.25	0.23	0.38	0.44	0.39	0.50	0.20	0.45	0.39
12	0.50	0.04	0.06	0.31	0.35	0.38	0.06	0.26	0.21	0.27	0.20	0.50	0.35	0.33
13	0.50	0.33	0.32	0.07	0.05	0.47	0.32	0.16	0.18	0.10	0.45	0.35	0.50	0.05
14	0.50	0.24	0.33	0.01	0.05	0.46	0.38	0.12	0.12	0.21	0.39	0.33	0.05	0.50
15	−1.00	−1.00	0.50	0.50	0.50	−1.00	0.50	0.50	0.50	0.50	0.50	0.50	0.50	0.50
16	0.50	0.03	0.03	0.27	0.40	0.39	0.03	0.20	0.25	0.30	0.23	0.10	0.30	0.30
17	0.50	0.34	0.31	0.06	0.11	0.47	0.32	0.16	0.17	0.13	0.43	0.35	0.02	0.02
18	0.50	0.32	0.33	0.05	0.05	0.47	0.33	0.16	0.16	0.20	0.45	0.35	0.01	0.03
19	0.50	0.18	0.12	0.20	0.26	0.44	0.33	0.03	0.02	0.12	0.38	0.21	0.19	0.17 −
20	−1.00	−1.00	−1.00	−1.00	−1.00	−1.00	−1.00	0.50	0.50	0.50	0.50	0.50	−1.00	−1.00 −
21	0.50	0.26	0.20	0.07	0.31	0.46	0.23	0.05	0.06	0.30	0.42	0.26	0.15	0.11
22	0.50	0.35	0.31	0.08	0.07	0.47	0.31	0.14	0.19	0.11	0.43	0.35	0.01	0.04
23	0.50	0.04	0.04	0.31	0.31	0.38	0.03	0.21	0.24	0.12	0.28	0.08	0.34	0.31 −
24	0.50	0.30	0.39	0.40	0.43	0.50	0.38	0.36	0.32	0.30	0.17	0.25	0.47	0.45
25	−1.00	−1.00	−1.00	−1.00	−1.00	−1.00	−1.00	0.50	0.50	0.50	0.50	−1.00	0.50	−1.00 −
26	0.50	0.22	0.17	0.14	0.34	0.45	0.20	0.01	0.0	0.17	0.40	0.26	0.18	0.14 −
27	0.50	0.31	0.28	0.02	0.21	0.47	0.30	0.12	0.16	0.04	0.47	0.35	0.04	0.02
28	−1.00	−1.00	−1.00	−1.00	−1.00	−1.00	−1.00	−1.00	−1.00	−1.00	−1.00	−1.00	−1.00	−1.00 −
29	−1.00	−1.00	−1.00	−1.00	−1.00	−1.00	−1.00	−1.00	−1.00	−1.00	−1.00	−1.00	−1.00	−1.00 −
30	−1.00	−1.00	−1.00	−1.00	−1.00	−1.00	−1.00	−1.00	−1.00	−1.00	−1.00	−1.00	−1.00	−1.00 −

TABLE 6 (Continued)

	17	18	19	20	21	22	23	24	25	26	27	28	29	30	
							Locational Type								
0	0.50	0.50	0.50	-1.00	0.50	0.50	0.50	0.50	-1.00	0.50	0.50	-1.00	-1.00	-1.00	1
3	0.34	0.32	0.18	-1.00	0.26	0.35	0.04	0.30	-1.00	0.22	0.31	-1.00	-1.00	-1.00	2
3	0.31	0.33	0.12	-1.00	0.20	0.31	0.04	0.39	-1.00	0.17	0.27	-1.00	-1.00	-1.00	3
7	0.06	0.05	0.20	-1.00	0.07	0.08	0.31	0.40	-1.00	0.14	0.02	-1.00	-1.00	-1.00	4
0	0.11	0.05	0.26	-1.00	0.31	0.07	0.31	0.43	-1.00	0.34	0.21	-1.00	-1.00	-1.00	5
9	0.47	0.47	0.44	-1.00	0.46	0.47	0.38	0.50	-1.00	0.45	0.47	-1.00	-1.00	-1.00	6
3	0.32	0.33	0.33	-1.00	0.23	0.31	0.03	0.38	-1.00	0.20	0.30	-1.00	-1.00	-1.00	7
0	0.16	0.16	0.03	0.50	0.05	0.14	0.21	0.36	0.50	0.01	0.12	-1.00	-1.00	-1.00	8
5	0.17	0.16	0.02	0.50	0.06	0.19	0.24	0.32	0.50	0.0	0.16	-1.00	-1.00	-1.00	9
0	0.13	0.20	0.12	0.50	0.30	0.11	0.12	0.30	0.50	0.17	0.04	-1.00	-1.00	-1.00	10
3	0.43	0.45	0.38	0.50	0.42	0.43	0.28	0.17	0.50	0.40	0.47	-1.00	-1.00	-1.00	11
0	0.35	0.35	0.21	0.50	0.26	0.35	0.08	0.25	-1.00	0.26	0.35	-1.00	-1.00	-1.00	12
0	0.02	0.01	0.19	-1.00	0.15	0.01	0.34	0.47	0.50	0.18	0.04	-1.00	-1.00	-1.00	13
0	0.02	0.03	0.18	-1.00	0.11	0.04	0.31	0.45	-1.00	0.14	0.02	-1.00	-1.00	-1.00	14
0	0.50	0.50	-1.00	-1.00	0.50	0.50	-1.00	0.50	-1.00	-1.00	0.50	-1.00	-1.00	-1.00	15
0	0.30	0.30	0.22	0.50	0.21	0.31	0.03	0.36	0.50	0.18	0.30	-1.00	-1.00	-1.00	16
0	0.50	0.0	0.19	-1.00	0.14	0.01	0.33	0.47	0.50	0.17	0.03	-1.00	-1.00	-1.00	17
0	0.0	0.50	0.19	0.50	0.13	0.01	0.36	0.47	0.50	0.16	0.02	-1.00	-1.00	-1.00	18
2	0.19	0.19	0.50	-1.00	0.03	0.22	0.21	0.38	-1.00	0.0	0.19	-1.00	-1.00	-1.00	19
0	-1.00	0.50	-1.00	0.50	0.50	-1.00	-1.00	-1.00	-1.00	0.50	-1.00	-1.00	-1.00	-1.00	20
1	0.14	0.13	0.03	0.50	0.50	0.14	0.26	0.45	0.50	0.04	0.10	-1.00	-1.00	-1.00	21
1	0.01	0.01	0.22	-1.00	0.14	0.50	0.33	0.46	-1.00	0.16	0.03	-1.00	-1.00	-1.00	22
3	0.33	0.36	0.21	-1.00	0.26	0.33	0.50	0.36	-1.00	0.23	0.32	-1.00	-1.00	-1.00	23
6	0.47	0.47	0.38	-1.00	0.45	0.46	0.36	0.50	-1.00	0.44	0.47	-1.00	-1.00	-1.00	24
0	0.50	0.50	-1.00	-1.00	0.50	-1.00	-1.00	-1.00	0.50	0.50	-1.00	-1.00	-1.00	-1.00	25
8	0.17	0.16	0.0	0.50	0.04	0.16	0.23	0.44	0.50	0.50	0.15	-1.00	-1.00	-1.00	26
0	0.03	0.02	0.19	-1.00	0.10	0.03	0.33	0.47	-1.00	0.15	0.50	-1.00	-1.00	-1.00	27
0	-1.00	-1.00	-1.00	-1.00	-1.00	-1.00	-1.00	-1.00	-1.00	-1.00	-1.00	0.50	-1.00	-1.00	28
0	-1.00	-1.00	-1.00	-1.00	-1.00	-1.00	-1.00	-1.00	-1.00	-1.00	-1.00	-1.00	0.50	-1.00	29
0	-1.00	-1.00	-1.00	-1.00	-1.00	-1.00	-1.00	-1.00	-1.00	-1.00	-1.00	-1.00	-1.00	0.50	30

TABLE 7

Perceived Similarity between Locational Types—Proximity Matrix, from Raw Data Matrix II
(Absolute Difference of Probabilities from 0.5 with −1 for Missing Data)

						Locational Type								
	1	2	3	4	5	6	7	8	9	10	11	12	13	14
1	0.50	0.50	0.50	0.50	0.50	−1.00	0.50	0.50	0.50	0.50	−1.00	−1.00	0.50	0.50 ·
2	0.50	0.50	0.50	0.31	0.50	0.50	0.07	0.25	0.50	0.50	0.50	−1.00	0.30	0.50 ·
3	0.50	0.50	0.50	0.50	0.50	0.50	0.50	0.0	0.25	0.50	0.50	0.50	0.36	0.40 ·
4	0.50	0.31	0.50	0.50	0.50	0.50	0.43	0.37	0.25	0.50	0.50	0.50	0.32	0.26 ·
5	0.50	0.50	0.50	0.50	0.50	0.50	0.50	0.50	0.50	0.20	0.50	0.50	0.45	0.45
6	1.00	0.50	0.50	0.50	0.50	0.50	0.50	0.50	0.50	0.50	−1.00	−1.00	0.50	0.50 ·
7	0.50	0.07	0.50	0.43	0.50	0.50	0.50	0.39	0.50	0.50	0.50	0.50	0.44	0.44 ·
8	0.50	0.25	0.0	0.37	0.50	0.50	0.39	0.50	0.30	0.50	0.50	0.50	0.0	0.38 ·
9	0.50	0.50	0.25	0.25	0.50	0.50	0.50	0.30	0.50	0.50	0.50	0.50	0.25	0.07 ·
10	0.50	0.50	0.50	0.50	0.20	0.50	0.50	0.50	0.50	0.50	0.50	0.50	0.50	0.30
11	−1.00	0.50	0.50	0.50	0.50	−1.00	0.50	0.50	0.50	0.50	0.50	−1.00	0.50	0.50 ·
12	−1.00	−1.00	0.50	0.50	0.50	−1.00	0.50	0.50	0.50	0.50	−1.00	0.50	0.50	0.50 ·
13	−1.00	0.30	0.36	0.32	0.45	0.50	0.44	0.0	0.25	0.50	0.50	0.50	0.50	0.24 ·
14	0.50	0.50	0.40	0.26	0.45	0.50	0.44	0.38	0.07	0.30	0.50	0.50	0.24	0.50
15	−1.00	−1.00	−1.00	−1.00	0.50	−1.00	−1.00	−1.00	−1.00	0.50	−1.00	−1.00	−1.00	0.50
16	0.50	0.25	0.50	0.44	0.50	0.50	0.25	0.50	0.50	0.50	0.50	−1.00	0.50	0.44
17	0.50	0.23	0.17	0.38	0.45	0.50	0.33	0.17	0.36	0.30	0.50	0.50	0.29	0.28
18	0.50	0.32	0.25	0.38	0.50	0.50	0.37	0.07	0.50	0.50	0.50	0.50	0.0	0.23
19	0.50	0.50	0.28	0.32	0.40	0.50	0.50	0.39	0.07	0.17	0.50	0.50	0.32	0.10
20	−1.00	−1.00	−1.00	−1.00	0.50	−1.00	−1.00	−1.00	−1.00	−1.00	−1.00	−1.00	−1.00	−1.00
21	0.50	−1.00	0.50	0.38	0.50	0.50	0.25	0.27	0.50	0.50	0.50	−1.00	0.35	0.50
22	0.50	0.50	0.07	0.31	0.38	0.50	0.30	0.17	0.17	0.07	0.50	0.50	0.28	0.31
23	0.50	0.06	0.10	0.50	0.46	0.50	0.17	0.33	0.39	0.33	0.50	0.50	0.26	0.43
24	−1.00	0.50	−1.00	0.50	0.50	−1.00	0.50	0.50	0.50	0.50	−1.00	−1.00	0.50	0.50
25	0.50	0.50	0.50	−1.00	0.50	0.50	0.50	0.50	−1.00	−1.00	0.50	0.50	0.50	−1.00
26	0.50	0.50	0.50	0.50	0.50	0.50	0.25	0.50	0.50	0.50	0.50	−1.00	0.50	0.44
27	0.50	0.50	0.21	0.50	0.50	0.50	0.23	0.41	0.50	0.50	0.50	0.50	0.44	0.26
28	−1.00	−1.00	−1.00	−1.00	0.50	−1.00	−1.00	−1.00	−1.00	−1.00	−1.00	−1.00	−1.00	−1.00
29	−1.00	−1.00	−1.00	−1.00	−1.00	−1.00	−1.00	−1.00	−1.00	−1.00	−1.00	−1.00	−1.00	−1.00
30	−1.00	−1.00	−1.00	−1.00	0.50	−1.00	−1.00	−1.00	−1.00	−1.00	−1.00	−1.00	−1.00	−1.00

TABLE 7 (Continued)

	Locational Type														
	17	18	19	20	21	22	23	24	25	26	27	28	29	30	
	0.50	0.50	0.50	−1.00	0.50	0.50	0.50	−1.00	0.50	0.50	0.50	−1.00	−1.00	−1.00	1
	0.23	0.32	0.50	−1.00	1.00	0.50	0.06	0.50	0.50	0.50	0.50	−1.00	−1.00	−1.00	2
	0.17	0.25	0.28	−1.00	0.50	0.07	0.10	−1.00	0.50	0.50	0.21	−1.00	−1.00	−1.00	3
	0.38	0.38	0.32	−1.00	0.38	0.31	0.50	0.50	−1.00	0.50	0.50	−1.00	−1.00	−1.00	4
	0.45	0.50	0.40	0.50	0.50	0.38	0.46	0.50	0.50	0.50	0.50	0.50	−1.00	0.50	5
	0.50	0.50	0.50	−1.00	0.50	0.50	0.50	−1.00	0.50	0.50	0.50	−1.00	−1.00	−1.00	6
	0.33	0.37	0.50	−1.00	0.25	0.30	0.17	0.50	0.50	0.25	0.23	−1.00	−1.00	−1.00	7
	0.17	0.07	0.39	−1.00	0.27	0.17	0.33	0.50	0.50	0.50	0.41	−1.00	−1.00	−1.00	8
	0.36	0.50	0.07	−1.00	0.50	0.17	0.39	0.50	−1.00	0.50	0.50	−1.00	−1.00	−1.00	9
	0.30	0.50	0.17	−1.00	0.50	0.07	0.33	0.50	−1.00	0.50	0.50	−1.00	−1.00	−1.00	10
	0.50	0.50	0.50	−1.00	0.50	0.50	0.50	−1.00	0.50	0.50	0.50	−1.00	−1.00	−1.00	11
	0.50	0.50	0.50	−1.00	1.00	0.50	0.50	−1.00	0.50	−1.00	0.50	−1.00	−1.00	−1.00	12
	0.29	0.0	0.32	−1.00	0.35	0.28	0.26	0.50	0.50	0.50	0.44	−1.00	−1.00	−1.00	13
	0.28	0.23	0.10	−1.00	0.50	0.31	0.43	0.50	−1.00	0.44	0.26	−1.00	−1.00	−1.00	14
	0.50	−1.00	0.50	−1.00	−1.00	0.50	0.50	−1.00	−1.00	−1.00	−1.00	−1.00	−1.00	−1.00	15
	0.50	0.50	0.50	−1.00	0.25	0.50	0.33	0.50	0.50	0.25	0.41	−1.00	−1.00	−1.00	16
	0.50	0.35	0.32	−1.00	0.29	0.04	0.19	0.50	0.50	0.50	0.33	−1.00	−1.00	−1.00	17
	0.35	0.50	0.40	−1.00	0.32	0.25	0.39	0.50	0.50	0.50	0.43	−1.00	−1.00	−1.00	18
	0.32	0.40	0.50	−1.00	0.50	0.14	0.40	0.50	0.50	0.50	0.50	−1.00	−1.00	−1.00	19
	−1.00	−1.00	−1.00	0.50	−1.00	−1.00	−1.00	−1.00	−1.00	−1.00	−1.00	−1.00	−1.00	−1.00	20
	0.29	0.32	0.50	−1.00	0.50	0.50	0.13	0.50	0.50	0.50	0.39	−1.00	−1.00	−1.00	21
	0.04	0.25	0.14	−1.00	0.50	0.50	0.20	0.50	0.50	0.50	0.08	−1.00	−1.00	−1.00	22
	0.19	0.39	0.40	−1.00	0.13	0.20	0.50	−1.00	0.50	0.33	0.05	−1.00	−1.00	−1.00	23
	0.50	0.50	0.50	−1.00	0.50	0.50	−1.00	0.50	0.50	−1.00	−1.00	−1.00	−1.00	−1.00	24
	0.50	0.50	0.50	−1.00	0.50	0.50	0.50	0.50	0.50	0.50	0.50	−1.00	−1.00	−1.00	25
	0.50	0.50	0.50	−1.00	0.50	0.50	0.33	−1.00	0.50	0.50	0.50	−1.00	−1.00	−1.00	26
	0.33	0.43	0.50	−1.00	0.39	0.08	0.05	−1.00	0.50	0.50	0.50	−1.00	−1.00	−1.00	27
	−1.00	−1.00	−1.00	−1.00	−1.00	−1.00	−1.00	−1.00	−1.00	−1.00	−1.00	0.50	−1.00	−1.00	28
	−1.00	−1.00	−1.00	−1.00	−1.00	−1.00	−1.00	−1.00	−1.00	−1.00	−1.00	−1.00	0.50	−1.00	29
	−1.00	−1.00	−1.00	−1.00	−1.00	−1.00	−1.00	−1.00	−1.00	−1.00	−1.00	−1.00	−1.00	0.50	30

TABLE 8
Coordinate Values for Locational Types, Proximity Matrix I

Rank	Locational Type	Coordinate Value
1	5	−0.597
2	22	−0.516
3	13	−0.510
4	27	−0.510
5	17	−0.505
6	14	−0.502
7	18	−0.497
8	10	−0.491
9	4	−0.474
10	21	−0.420
11	19	−0.415
12	26	−0.408
13	9	−0.408
14	8	−0.403
15	3	−0.331
16	2	−0.318
17	16	−0.310
18	23	−0.309
19	12	−0.293
20	7	−0.281
21	24	−0.157
22	11	−0.143
23	6	0.173
24	28	1.174
25	29	1.174
26	30	1.174
27	15	1.242
28	1	1.267
29	25	1.274
30	20	1.323

NOTE: Calculated from Torgerson multidimensional scaling program (TORSCA).

Muth [12], and Wingo [21] models.[1] Overall, the most-preferred indifference curve would show both low density and relative closeness to the center of the city. Less-preferred indifference curves would be to the right of that line. (See Figure 2.) The second set of proximity data produces a similar pattern of indifference curves (Figure 3). However, a close inspection of the maps suggest that the indifference surface is not simply declining away from locational types 1 and 6. The maps show that there are comparable values at the low density-short distance combinations and

TABLE 9
Coordinate Values for Locational Types, Proximity Matrix II

Rank	Locational Type	Coordinate Value
1	3	−1.024
2	22	−0.940
3	9	−0.922
4	19	−0.897
5	17	−0.797
6	14	−0.752
7	13	−0.701
8	8	−0.694
9	18	−0.621
10	23	−0.497
11	4	−0.461
12	27	−0.400
13	10	−0.281
14	5	−0.206
15	7	−0.075
16	21	−0.006
17	2	0.001
18	16	0.194
19	25	0.342
20	26	0.406
21	12	0.569
22	6	0.652
23	11	0.652
24	1	0.652
25	24	0.736
26	15	0.952
27	29	1.011
28	20	1.036
29	28	1.036
30	30	1.036

NOTE: Calculated from Torgerson multidimensional scaling program (TORSCA).

at the high density-long distance combinations. If the shaded areas were included, we could imagine a curved surface with the hollow of the surface running diagonally from the southeast to the northwest. This result is more pronounced in the second proximity matrix, where constructing the isopleth leads to a suggested surface with a definite concave character. It is important to speculate, at least, on the possibility of repeated indifference curves.

When we recognize that the highest-ranked locational type is 5 for proximity matrix I and 3 for proximity matrix II, we can

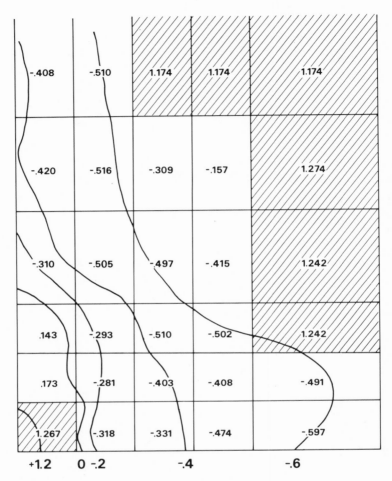

Figure 2. Isopleth lines for the scale values derived from proximity matrix I.

suggest an alternative interpretation (incorporating dual indifference curves) of the preference map. With a maximum indifference curve of − .6 in the first data set and − 1.0 in the second data set, we might suggest an elongated tent set up over the highest indifference curve, implying two sets of indifference curves on either side of the "highest" indifference curve. Although this may appear illogical, it could very well reflect the fact that there is a highly desired combination of distances and densities and that there is a

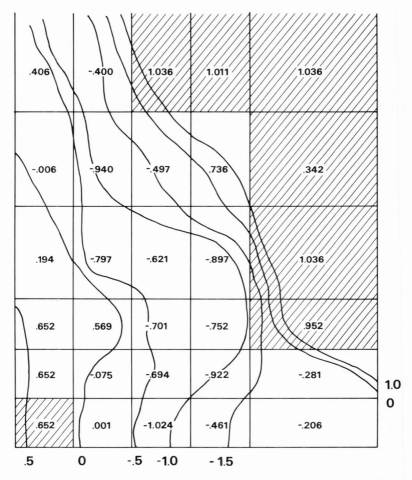

Figure 3. Isopleth lines for the scale values derived from proximity matrix II.

shading away from this in directions either toward the city (which would probably be an older housing stock) or toward the rural sections of the city (which would necessarily have fewer amenities). This speculation of an elongated tent set up over the highest indifference curve, implying two sets of indifference curves on either side, has little to support it as yet, so an alternative explanation is required.

The shaded areas on both maps represent locational types that either were not chosen or were chosen by only one or two migrants. Thus, the alternative explanation of the isopleth results is that they are to some extent caused by peculiarities in the data.[2] At the same time, it is quite clear that locational type 5 is a highly preferred type (Tables 4 and 5). The probabilities that type 5 will be chosen over other types are uniformly high. Somewhat similar results can be seen for locational type 22. These fundamental aspects of the scale do not change. It is not unreasonable to suggest that the most-preferred indifference curve is at either −0.6 or −1.0, respectively, and that less-desired indifference curves are to the left of that curve. In such an interpretation, the data are insufficient for analyses beyond these points. However, only an increase in the number of revealed preferences will answer some of these speculations. The data for proximity matrix I included 450 choices and for proximity matrix II, only 369 cases. (The number of cases is smaller for the second set because there were not journey-to-work data for all respondents.) When there is a small number of cases and a reasonably large number of locational types, sampling error will be large. By increasing the sampling size it will be possible to verify the speculative results developed so far.

CONCLUSION

The intent of this paper was to validate the revealed preference methodology for a situation other than that of consumer behavior. Although the development of a unidimensional scale was successful, the attempt to map that scale as an indifference surface was less satisfactory.[3] Although a different scale and perhaps more expressive surface would be possible with a larger data set, it does seem that the methodology could be applied usefully in an intraurban migration context to uncover the preferences of a population of intraurban movers. The most important implication to be derived from the work here is "the potential to separate preferences and opportunities" [16, p. 223] that is fundamental to an understanding of the flows of population in the city. Although some of the results are certainly speculative, they indicate the broad direction for a continuing and more detailed application of revealed preference analysis to intraurban migration choices.

NOTES

1. At first consideration, it might appear that the results of the scaling are merely capturing the distribution of urban densities at set distances from the CBD and so the basic urban structure of the city. However, though the aggregate pattern of densities within the metropolitan area is one of decline with distance from the CBD, there is considerable variation in population density per acre in any subsector of the city.

2. In a personal communication (August 16, 1978), Gerard Rushton suggested the significance of missing data. His comments bear repeating. "TORSCA substitutes the mean of the data matrix for missing data values and therefore those locational types 20, 25, 28, 29, and 30 would have had the mean substituted very frequently. That is consistent with the results you get. I would recommend that you reduce your proximity matrix by removing the columns and corresponding rows where there are large numbers of missing data. You are making inferences from empirical data and if no one in your sample went or could go to those types, you have no basis for inferring peoples' preferences about them. Two principles are involved. One is that you can only calibrate a preference function within the domain of experience and the second is more subtle and goes back to Cattells' 1890's dictum about the method of paired comparisons —*equally often noticed differences are equal —unless always or never noticed.* It means that if one location type is always chosen over say five others, we cannot infer that the preferential distance is equal between it and the other five. Likewise, if five types are never chosen over a given type, we can't infer that the preferential distance is equal." The removal of types 20, 25, 28, 29, and 30 did not, however, materially change the results reported in Figure 2; for Figure 3 the results were less easily interpreted.

3. Recently, there has been a spirited debate about the role and usefulness of the revealed preference methodology (MacLennan and Williams, 1979, and Timmermans and Rushton, 1979). The critique suggests that the application of an aspatial theory (consumer preference) in a spatial context may very well invalidate it. Thus, in addition to problems created by complex data, it is possible that the difficulty of achieving consistent results may be in part due to the translation of the methodology to a spatial context.

MacLennan, D., and Williams, N. J., 1979. "Revealed Space Preference Theory—A Cautionary Note," *Tijdschrift voor Economische en Sociale Geografie,* 70, 307-9.

Timmermans, H., and Rushton, G., 1979. "Revealed Space Preference Theory—A Rejoinder," *Tijdschrift voor Economische en Sociale Geografie,* 70, 309-12.

REFERENCES

1. Alonso, W., 1964. *Location and Land Use.* Cambridge, Mass.: Harvard University Press.
2. Barrett, F., 1973. *Residential Search Behavior.* Toronto: York University Research Monographs.
3. Burnett, P., 1973. "The Dimensions of Alternatives in Spatial Choice Processes," *Geographical Analysis,* 5, 181-204.

4. Clark, W. A. V., and Rushton, G., 1970. "Models of Intra-Urban Consumer Behavior and Their Implications for Central Place Theory," *Economic Geography*, 46, 486-97.

5. Demko, D., 1974. "Cognition of Southern Ontario Cities in a Potential Migration Context," *Economic Geography*, 50, 20-34.

6. Festinger, L., 1957. *A Theory of Cognitive Dissonance*. Stanford, Calif.: Stanford University Press.

7. Girt, J. L., 1976. "Some Extensions to Rushton's Spatial Preference Scaling Model," *Geographical Analysis*, 8, 137-52.

8. Golledge, R. G., and Rushton, G., 1972. *Multidimensional Scaling: Review and Applications*. Washington, D.C.: Association of American Geographers, Commission on College Geography, Technical Paper #10.

9. Gould, P., 1967. "Structuring Information on Spatio-Temporal Preferences," *Journal of Regional Science*, 2, 259-74.

10. Lieber, S. R., 1976. "A Comparison of Metric and Nonmetric Scaling Models in Preference Research." In R. G. Golledge and G. Rushton (eds.), *Spatial Choice and Spatial Behavior: Geographic Essays on the Analysis of Preferences and Perceptions*. Columbus, Ohio: Ohio State University Press.

11. Louviere, J. J., 1976. "Information Processing Theory and Functional Form in Spatial Behavior." In R. G. Golledge and G. Rushton (eds.), *Spatial Choice and Spatial Behavior: Geographic Essays on the Analysis of Preferences and Perceptions*. Columbus, Ohio: Ohio State University Press, 148-86.

12. Muth, R. F., 1969. *Cities and Housing*. Chicago: University of Chicago Press.

13. Palm, R., 1976. *Urban Social Geography from the Perspective of the Real Estate Salesman*. Berkeley, Calif.: University of California, Berkeley Center for Real Estate and Urban Economics.

14. Pleeter, S., 1976. "Residential-Location Theory: Some Modifications," *Geographical Analysis*, 6, 369-75.

15. Rossi, P., 1955. *Why Families Move*. New York: Free Press.

16. Rushton, G., 1969. "The Scaling of Locational Preferences." In K. R. Cox and R. G. Golledge (eds.), *Behavioral Problems in Geography: A Symposium*. Evanston, Ill.: Department of Geography, Northwestern University, Studies in Geography #17.

17. Rushton, G., 1969. "Analysis of Spatial Behavior by Revealed Space Preference," *Annals, Association of American Geographers*, 59, 391-400.

18. Stegman, M., 1969. "Accessibility Models and Residential Location," *Journal of the American Institute of Planners*, 35, 22-34.

19. White, R. W., 1976. "A Generalization of the Utility Theory Approach to the Problem of Spatial Interaction," *Geographical Analysis*, 8, 39-46.

20. Wicker, A. W., 1969. "Attitudes Versus Actions: The Relationship of Verbal and Overt Behavioral Responses to Attitude Objects," *Journal of Social Issues*, 28, 41-78.

21. Wingo, L., 1961. *Transportation and Urban Land*. Baltimore: John Hopkins Press.

Chapter 2.4 Data Problems and the Application of Conjoint Measurement to Recurrent Urban Travel

Patricia K. Burnett

Over the past few years, some interest has been shown in the decomposition of space preference functions [20, 21, 25, 27]. It is desirable to recover the part-worths of different attributes of spatial alternatives in the choice process. One rationale for this interest is that it is necessary for practical purposes to predict the degree to which an attribute must be altered in order to achieve a desired spatial choice effect. It is necessary, for example, to be able to predict the spatial choice effects of a deliberate alteration of the attributes of shopping destinations, such as price, quality of goods, and convenience. It is also necessary at times to predict the effects of uncontrolled alterations in some attributes of destinations. A good example is the alteration of the convenience of shopping destinations as a by-product of major investments in transportation facilities. Further, because of the combinatorial quality of data-collection designs, relatively few attributes of alternatives can be presented for judgment in a choice experiment. It is useful to be able to predict responses for many combinations of the attributes of alternatives where only the part-worths of a few combinations are known. The development of conjoint measurement

The author gratefully acknowledges the support of the Department of Transportation (contract DOT-OS-30093) for the Council for Advanced Transportation Studies at the University of Texas at Austin.

as a multidimensional scaling technique permits the recovery of part-worths of attributes for all of these applications [14, 15, 16, 18, 19, 22, 28, 31].

Despite the apparent usefulness of conjoint analysis, few geographers have employed it so far. Brummel and Harman [3, pp. 49-52], Lieber [20], and Rushton [25] present introductory discussions of conjoint measurement for spatial choice, but they only examine simple laboratory applications. In addition, there are alternative ways of approaching the part-worths of the attributes of alternatives via trade-off functions, such as those suggested by Louviere, Wilson, and Piccolo [21] and Stutz [27]. These methods, however, have rarely been shown to be capable of predicting responses to altered spatial alternatives from the utilities of attributes of existing ones.

The use of conjoint measurement for spatial choice, therefore, has some support, but it remains largely unexplored. This situation is exacerbated by the fact that attributes of spatial alternatives are still commonly represented by surrogates such as size, distance, and quality [9, 23, 24, 25, 29] or are neglected altogether [8]; thus, changes in choice behavior following changes in the variables that control choice cannot be directly examined. Consequently, it is the objective of this paper to detail problems and their resolution in a pilot real-world application of conjoint measurement to spatial choice. It is particularly germane to consider how both new and existing spatial choice behavior might be predicted by the technique. The use of multidimensional scaling techniques has not yet passed from descriptions to forecasts of spatial behavior [as, for example, in 2, 4, 5, 10, and 20]. The increasingly common use of conjoint analysis in analogous marketing and brand-choice situations appears to demonstrate the effectiveness of the technique for forecasting purposes [6, 7, 11, 12, 13, 16, 30].

THE CONJOINT MEASUREMENT MODEL
FOR SPATIAL CHOICE

Familiar spatial choice behavior includes long-run migration decisions and those short-run destination selections of recurrent trip making. In the interest of simplicity, this paper focuses on travel comprising the selection of one out of n destinations on each of a series of trips from a single base.[1] The paradigmatic example is shopping behavior, though social and recreational travel and residential search are also embraced. The conjoint measurement approach is outlined for recurrent travel, largely following Johnson [15, 16].

Trade-offs between attributes of alternatives govern travel behavior. For conjoint measurement, different levels of attributes act as stimuli. For example, different levels of convenience of alternatives (close, distant) might influence travel. In the choice decision, combinations of stimuli are traded against each other; a higher-priced, nearby store may be more acceptable than a cheaper, more distant store when price is traded off against distance [20]. The general goal of conjoint analysis is "the decomposition of complex phenomena into sets of basic factors according to some specified rules of combination" [31, p. 69]. In modeling spatial choice, "the complex phenomena" are the preference rankings assigned to different spatial stimuli combinations. An example of such a preference ranking is shown in Table 1. The conjoint measurement problem is to obtain a measurement of each of the individual stimuli such that the combinations of measures account for the rank order of the preferences. The combination rule may be of an additive or a multiplicative variety [20].

One appropriate conjoint measurement model may be described as follows. Consider the matrix X with nine cells representing combinations of three levels one with another. The matrix X is of order $n \times p$, where in the sample case, $n = 9$ "objects" (cells describing combinations of stimuli and $p = 6$ "independent" variables (levels of attributes). Let there be m individuals, and let the vector Y of length n contain an individual's preference ratings for the n combinations; there will be m matrices and vectors, one for each individual. Now consider a single unknown vector W of length p containing weights. Let $XW = Z$ for each individual (the additive model) or $X \log_e W = Z$ (the multiplicative model). Then, the conjoint measurement problem is finding one W so that the elements of each Z and Y have as nearly similar rank orders as possible. The goodness-of-fit statistic is:

$$\theta^2 = \frac{\sum\limits_{k}^{m}\sum\limits_{ij}\delta_{ijk}(z_{ik} - z_{jk})^2}{\sum\limits_{k}^{m}\sum\limits_{ij}(z_{ik} - z_{jk})^2}$$

where:

$$\delta_{ijk} = \begin{cases} 1 \text{ if sign } (z_{ik} - z_{jk}) \neq \text{sign } (y_{ik} - y_{jk}) \\ 0 \text{ otherwise} \end{cases}$$

TABLE 1
Preference Ranking of Spatial Alternatives for Recurrent Travel

Distance	Price		
	High	Medium	Low
Close	5	2	1
Middle	6	4	3
Far	9	8	7

(This statistic is discussed by Johnson [14].) The statistic has limiting values of 0 when the elements of all Z and Y are of exactly similar rank and a value of 1.0 "if the rank order of predictions are exactly the opposite of the input data" [15, p. 3]. When the recovered vector W is such that $\Theta^2 = O$, then the elements of the single vector W can be considered to be the utilities of the group of m individuals for each of the levels of the p destination attributes, following Green and Wind [13, pp. 109-10], among others. This is because multiplication or addition of the weights for any attribute levels yields the preference rank of the level combinations, so that it is "natural" to interpret the weights as group utilities.[2]

The group utilities for attribute levels comprise the desired part-worths of each destination attribute. There is a group utility for all possible combinations of levels of destination attributes, whether or not there actually exists a destination described by each combination. Thus, conjoint analysis can predict the utility of changed levels of destination attributes and of new destinations with combinations of attributes not hitherto in existence. Once the utilities are calculated, it is but a short step to the estimation of choice probabilities, as will be shown. The only caveat is that new or altered destinations must be defined by some of those combinations of attribute levels that are used in analysis.[3]

There are three basic properties of the model that should be mentioned, since they create special problems in the application of conjoint measurement to spatial choice. These are the assumption that all relevant alternatives and the attributes controlling choice are known, the assumption that attributes are independent, and the question that asks whether the combination rule is of a multiplicative, additive, difference, or some other variety [3, pp. 57-62]. The special difficulties that these assumptions create will be further discussed below.

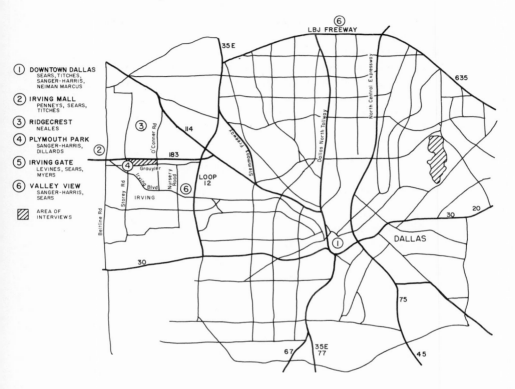

Figure 1. The study area.

A case study approach is adopted to explore problems of the application of the conjoint measurement model to spatial choice. The intent is to highlight some of the problems and methods for their resolution. The shopping travel behavior of residents of Irving, Texas, was selected for examination. Irving was a city of 116,032 persons in 1970, and it is located just outside the city limits of Dallas, Texas (Figure 1).

Before conjoint measurement can be applied, the relevant shopping alternatives and their attributes must be defined. This means that, ideally, in order to use the technique a two-phase survey approach must be adopted. The first phase involves the delineation of respondents' possible destinations and the destination attributes to which they respond; the second involves the application of conjoint measurement itself. The need for two surveys means that applications of the technique are more expensive than other multidimensional scaling analyses—for example, the one-phase use of TORSCA to describe how shopping travel flows arise from perceptions of, and preferences for, destination alternatives [4]. However, conjoint analysis yields more information about responses to new as well as existing destination alternatives and about how utility and preference functions arise; thus, it remains "one of the most promising avenues of scaling for behavioral geography" [3, p. 51].

THE PHASE-ONE SURVEY

The Destination Choice Set Problem

General Problem. In choice experiments in the laboratory, there is a single a priori set of alternatives to be presented to every subject. [For the choice paradigm, see 1, pp. 137-39.] By contrast, in the spatial choice situation, it is difficult to define, a priori, that set of destination alternatives that each individual uses for selection. Presumably, the choice set comprises that set of destinations with which a respondent is familiar for a given trip purpose. However, there needs to be some specification of the degree of familiarity that is required for a destination to be included in the choice set. For an alternative to lie within the domain of analysis, does it need to have been used, or heard about, or does some other criterion of familiarity need to be employed? There has been little exploration of this question in the literature.

Other questions are also raised concerning the choice set. For example, the set might vary between individuals, depending on their demographic characteristics. In addition, there is no reason to suppose that the choice set is stable over time: the choice set might vary over successive selections as the individual learns more about the environment.

One strategy that Rushton [23, 24] and others [17] have used to resolve some of these problems is classifying alternatives in some way so that all classes are available to each individual. For example, one convenient scheme for shopping travel is classifying shopping center destinations according to distance and size surrogates following central place theory. However, this is open to the now well-known criticism that the surrogates defining classes in the destination choice set do not represent the actual stimuli to which consumers respond. Consequently, the choice set may not contain alternatives defined in a way that is relevant to choice behavior

[3, pp. 79-81]. This is of importance when one wants to predict precisely what alterations in alternatives will be associated with particular spatial choice effects, as in the present instance.

Another strategy, therefore, appears to be preferable. This is to select a group of respondents from adjacent blocks within a well-established urban neighborhood. This strategy, like the first, represents an attempt to minimize discrepancies among individuals in the set of destinations for a specific trip purpose. Individuals from adjacent blocks are likely to have similar demographic characteristics, and their choice set should be stable when the neighborhood is an older, well-established one. The case study investigated the ability of this solution to define the choice set.

Case Study Resolution. In the first-phase survey, the blocks shown in Figure 1 were selected as the case study area. All homes within the area were approached, with one callback, yielding forty-eight respondents for the phase-one survey. Visits to department stores were selected as the trip purpose for investigation; others have covered urban area shopping center selection [26], grocery store selection [17], and women's wear selection [4] via multi-dimensional scaling. A list of fifteen possible department stores was compiled from directories and a field survey (Figure 1). Each respondent was asked whether he or she had heard of each store (one operational definition of familiarity) and to list others that he or she knew but that had not been included. An analysis of the responses is given in Table 2.

The results show that the percentage of respondents aware of a store generally showed a small variation about a mode of 77%. This seems to reveal only small discrepancies in individuals' choice sets. Such a conclusion is also supported by the additional stores that were mentioned. These ranged in number from zero to four for a respondent; however, only 35% of the respondents listed stores other than those on the original list, and the additional stores were different in each case. It is noteworthy that all individuals traveled from home to use department stores, so that different trip origins did not influence the choice set. It appears that, while selecting respondents from a single urban neighborhood does not altogether resolve problems of choice set definition, especially for aggregative modeling, it is a reasonable pilot-study strategy for delineating the set of alternatives used by a group of individuals for a given shopping purpose. Consequently, the fifteen stores listed in Table 2 were retained for later analysis by conjoint measurement on the

TABLE 2
Percentage of Respondents Aware of Destination

Destination	Percentage
1	76.92
2	76.92
3	76.92
4	78.85
5	98.08
6	98.08
7	96.15
8	42.31
9	88.46
10	88.46
11	78.85
12	78.85
13	75.00
14	67.31
15	65.38

grounds that they comprise the choice set of department stores for consumers in the sampled neighborhood.[4]

Problems of Attribute (Stimuli) Definition

General Problem. To apply conjoint analysis, not only does there need to be an a priori definition of the choice set, there also needs to be an a priori definition of attributes describing the set. This is because levels of attributes comprise the stimuli to which individuals respond in the choice experiment, as noted in the model above. In laboratory situations and in many social science applications of choice theory, relevant attributes of alternatives are well defined. This is not so in spatial choice behavior.

There are two widespread solutions to this problem. First, a checklist of attributes that seem applicable may be presented to respondents, and some measure of the importance of each attribute may be elicited (for example, Stutz's [27] checklist of factors affecting destination patronage). This simple method, when used alone, suffers from the fact that the checklist may not contain all the attributes to which individuals respond, and the attributes may not be defined in the manner in which they are used (for example, value for money may be used instead of price). Moreover, of course, individuals may not respond truthfully or accurately, the source of response bias. In order to avoid these difficulties, nonmetric multidimensional scaling techniques are now widely used. Through applying multidimensional unfolding procedures, for example, attributes used to assess destinations can be recovered from a simple ranking of choice set members in order of preference. The naming of recovered destination attributes is, however, a well-known problem for which no satisfactory solution exists [3, pp. 55-56]. There is no objective means of interpreting recovered attributes without recourse to exogenous evidence on the properties of destinations [4].

Case Study Resolution of Attribute Definition. A combination of both solutions was used in the first phase of the case study. This represents an attempt to use techniques that are applicable in a short field interview and that allow cross-checking of results.

The forty-eight respondents in the phase-one survey were asked to check those listed attributes of alternatives that were important in their selection of a department store. They were also asked to rank order all fifteen of the stores in the choice set in the order of their preference; the preference data were then scaled by using TORSCA-9 for analysis. The best solution was a four-dimensional one in Minkowski metric two.

The checklists and MDS solution were used in the naming of relevant destination attributes in the following way. First, calculations were made of the percentages of respondents who checked each attribute on the list. The results of these calculations are shown in Table 3. They indicated that price level, variety of merchandise, time to store, and prestige were the main factors in store choice. Consequently, each of the fifteen stores was rated from 1 to 7 on each of these attributes on a field survey, with 1 representing the bottom score. The store ratings for each attribute were compared with the coordinate values of stores on each dimension obtained from the multidimensional scaling analysis. A correlation statistically significant at the 5% level suggested that a dimension could be interpreted as the attribute with which it was being compared. The results of the naming procedure are shown in Figures 2 to 4.

This procedure does not, of course, guarantee that the attributes describing destinations have been identified. However, a high degree of confidence can be placed in the results. Both the checklist and the field survey identified attributes with a high degree of correlation with dimensions derived from multidimensional scaling.

The results are also very pleasing from the point of view of identifying the stimuli required for conjoint analysis. Each recovered

TABLE 3
Percentage of Respondents Checking Attributes as Important

Attribute	Percentage
Price	*88.46
Variety of merchandise in store	*67.31
Number of adjacent stores	36.54
Parking	48.08
Time taken to store	*59.61
Distance to store	51.92
Safety of shopping environment	26.92
Design, layout of store	17.31
Services	51.92
Quality of merchandise	19.23
Prestige of store	*75.00
Design, layout of shopping center	15.38
Advertising	15.38

*One of four highest-ranking attributes.

dimension has a dispersion of destinations along it. Several different levels of each attribute thus act as stimuli in the case study situation, as is required by the conjoint analysis model.

THE PHASE-TWO SURVEY

The phase-two survey comprised the application of conjoint analysis to determine the part-utilities attached to each level of destination attribute, together with the utility of the whole bundle of attribute levels defining a destination. The problems in this phase of the survey occur not only in real-world spatial choice, but also in laboratory and other choice situations in general. Their resolution is, therefore, of widespread concern.

The Complexity of Data-Gathering Procedures

It has been noted that conjoint analysis is used to analyze the preference rankings assigned to different spatial stimuli combinations (Table 1). The procedure is designed to reveal the utility of each level of a destination attribute relative to the utility of each level of each other attribute. In order to perform the analysis, a matrix like the one in Table 1 is required for each possible combination of attributes; if there are *n* attributes of interest, there will

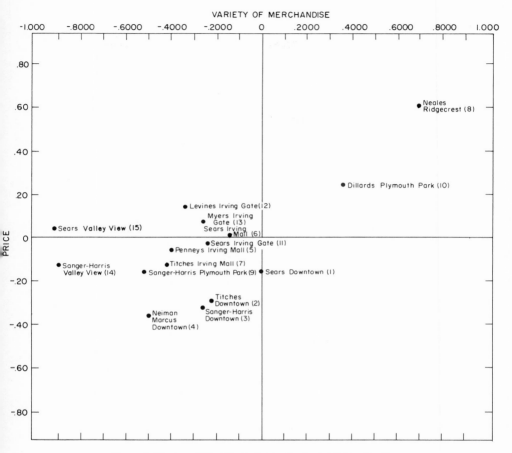

VARIETY OF MERCHANDISE

Figure 2. Store configuration: dimension 1 and dimension 2.

be $n(n - 1)/2$ matrices. A problem of complexity in data gathering thus arises when *n* becomes large.

There is, however, a good reason for believing that this might not be a difficulty in spatial choice situations. Previous studies (for example, [4]) have shown that the number of attributes that are used to discriminate alternatives is small. In the present instance, $n = 4$ (variety of merchandise, price, convenience, prestige), so that only six matrices needed to be presented to subjects. Each attribute was divided into three levels, and respondents were, therefore, required to fill in six matrices of the type shown in Table 1.

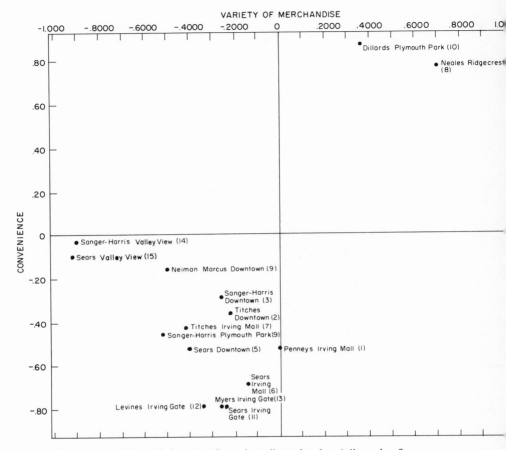

Figure 3. Store configuration: dimension 1 and dimension 3.

There were 100 respondents from whom these data were gathered on the phase-two survey. They were sampled from the same area as phase-one respondents (Figure 1). This enabled the necessary (though questionable) assumption to be made that the subjects used the same attributes to discriminate alternatives that the sample for the phase-one survey did.

The Independence of Attributes Assumption

The application of conjoint analysis requires that the attributes used be independent of each other; that is, "there should be no

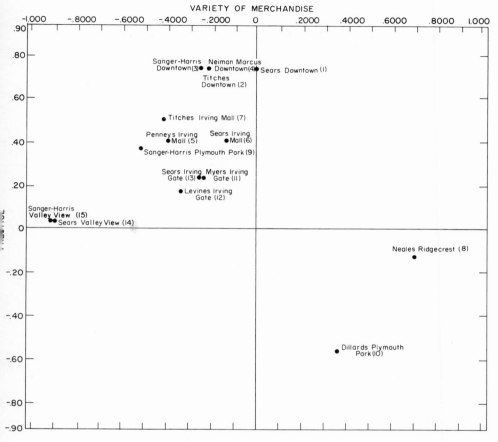

Figure 4. Store configuration: dimension 1 and dimension 4.

interaction effects between attributes" [3, p. 51] and "the attributes must all be nonredundant, or, more accurately, they all must be equally redundant" [16, p. 124]. In a spatial choice situation, it is by no means self-evident that the independence-of-attributes assumption is "tenable under ordinary circumstances. However, "if interactions do exist in a special set of data, they will be indicated by unfavorable values" of the goodness of fit of the statistic [16, p. 124]. The analysis of the respondents' six preference matrices was, therefore, carried out, with Johnson's trade-off analysis algorithm being used. This algorithm finds the solution to the conjoint

TABLE 4
θ Statistics for the Additive Model

Matrix Number	Attributes	θ
1	Variety Price	.21916
2	Variety Convenience	.28299
3	Variety Prestige	.36280
4	Price Convenience	.28596
5	Price Prestige	.35656
6	Convenience Prestige	.31929

measurement problem outlined in the first section of the paper. Table 4 shows the six θ statistics obtained by the analysis, one for each of the six matrices. In the case study, low values of the θ statistic for the additive model indicate that little interaction is present. Accordingly, attributes of alternatives seem independent enough to yield meaningful results.

The output of the conjoint analysis program is shown in Table 5. This table contains the weights (utilities) derived for each level of

TABLE 5
Weights (Utilities) for the Additive Model

Matrix	Level of Attribute		
	High	Medium	Low
Variety × Price	.41	.04	.19
Variety × Convenience	.41	.004	.34
Variety × Prestige	.63	.03	.39
Price × Variety	.10	.03	.25
Price × Convenience	.17	.05	.40
Price × Prestige	.34	.03	.58
Convenience × Variety	.21	.01	.15
Convenience × Price	.14	.00	.15
Convenience × Prestige	.56	.01	.41
Prestige × Variety	.09	.02	.03
Prestige × Price	.01	.002	.003
Prestige × Convenience	.05	.01	.01

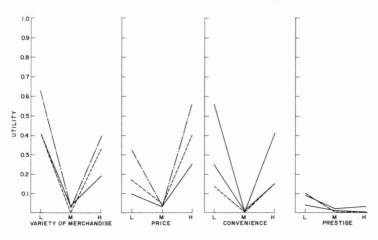

Figure 5. Utilities for attribute levels.

each attribute, with one set of weights derived for each of the matrices in which any given attribute appears. Figure 5 contains plots of the utilities for the levels of attributes that are portrayed in Table 4.

Interpretation of Results

The Assumption of Additivity. One assumption underlies the presentation of results thus far. This assumption is that the additive conjoint model is preferable to the multiplicative model. In general, reasons for the assumption in practice of either the additive or the multiplicative combination rule are rarely discussed in the literature on the application of the conjoint measurement model. [See, for example, 16, 28.] Since each model yields different results, it is worthwhile to examine briefly the criteria for the use of one or the other form.

Obviously, one criterion that can be used is the goodness-of-fit statistic, θ. For the present instance, the θ statistic for the multiplicative model was higher than the θ for the additive model, for all six matrices in the experiment. This indicates a better fit of the additive model to the original preference data and justifies its use.

Another criterion could be the kind of relationship that may be anticipated a priori among the variables. For example, in the additive model, it is implied that the total utility of a given destination is the sum of the utilities for each level of the attributes that describe it. This implies that, if one of the utilities-of-destination attributes is zero, then the destination will have a utility that is a sum of the utilities of the remaining attribute levels. This seems reasonable. In contrast, for the multiplicative model, the multiplication of the utilities of each attribute level gives the overall utility of a destination. This implies the unreasonable result that, if only one of the utilities is zero, the total utility of the destination is zero. Accordingly, the additive rather than the multiplicative model seems preferable for most applications, despite the fact that the multiplicative model is recommended for widespread use [16]. This is a second reason for concentrating here on the results of the additive model.

Computational and Interpretive Problems. As can be seen from Table 5, each matrix in which an attribute occurs produces a different set of utilities for levels of that attribute. A question arises as to which set of utilities should be used for the purpose of further computation and interpretation. This question is not discussed in the literature, although the resolution of the problem is of some importance. It can be seen from Figure 5 and Table 5 that the values for each attribute level are similar in each of the three matrices in which they occur. Consequently, the variation from matrix to matrix may reflect deviations in the utility value of an attribute level about some overall mean. The mean value of the utility of each attribute level was therefore calculated. These mean values are shown in Table 6 and Figure 6. Each attribute in these illustrations has the same pattern of utility values. Peak utility is found at the low level and the high level of each attribute. This apparently reflects the fact that, for the case study respondents, there is most preference for either department stores that are very convenient or are low priced but that have a low variety of merchandise and low prestige or for department stores that are less convenient or are

TABLE 6
Average Utilities for the Additive Model

Attribute	Level of Attribute		
	High	Medium	Low
Variety	.48	.02	.31
Price	.20	.04	.41
Convenience	.30	.01	.24
Prestige	.05	.01	.01

high priced but that have great variety and high prestige. This is a reasonable result.[5] Consequently, the mean utility for each attribute level can be used for further analysis.

In particular, the overall mean utility for all attributes of each store in the case study can be calculated. The sum of the mean utility for each attribute level describing a destination can be found. The problem here is determining which levels of attributes define each destination in order to carry out the computation. Perceived levels of attributes determine behavior (for example, the perceived convenience of a destination influences its selection). This suggests that configurations of destinations like those in Figures 2, 3, and 4 can be used to define whether choice alternatives rank high, medium, or low on each attribute. In the case study, therefore, the levels of each attribute that describes a store were derived from an analysis of Figures 2, 3, and 4. Table 7 contains the results of this analysis and of the summing of the utilities of each attribute level describing a store which are used to compute overall store utility.

The store utilities can now be easily related to choice probabilities; thus, a model of destination choice can be generated by the conjoint analysis procedures. The utility of a destination is reflected in the probability with which it will be chosen. Let the aggregate utility, U, of all destinations be equal to the sum of the utilities of individual destinations, U_j. Then, a destination alternative

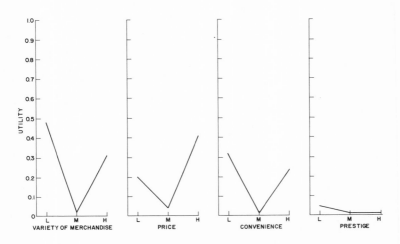

Figure 6. Average utilities for attribute levels.

TABLE 7
Store Utilities

Store	Variety	Attribute Level* Price	Convenience	Prestige	Total Utility†
1	M	H	M	H	0.28
2	M	H	M	H	0.28
3	M	H	M	H	0.28
4	H	H	M	H	0.74
5	M	M	M	M	0.08
6	M	M	L	M	0.31
7	H	M	M	H	0.58
8	L	L	H	L	1.03
9	H	H	M	M	0.70
10	L	L	H	L	1.03
11	M	M	L	M	0.31
12	M	M	L	M	0.31
13	M	M	L	M	0.31
14	H	M	M	M	0.54
15	H	M	M	M	0.54

*L = low, M = medium, and H = high.
†Utilities are calculated from Table 6.

will have a choice probability of U_i/U. Table 8 contains the result of calculating the choice probabilities for the case study department stores. There is some difficulty with the interpretation of the choice probabilities derived in this fashion. Should they be interpreted as the relative frequency with which an entire group of individuals will select an alternative, so that, for example, a destination choice probability of .50 means that a group of subjects will allocate 50% of their aggregate number of trips to that alternative? Or should choice probabilities be interpreted as the relative frequency with which each individual selects an alternative? Since the conjoint analysis model in this paper yields only *group* utilities for different levels of destination attributes, the first interpretation seems preferable.

Finally, the results so far can be used to predict the utility (and hence the choice probability) of new alternatives in the choice set. For example, there could be added to the existing department stores in the case study a new store with an image of low price, low prestige, high convenience, and high variety of merchandise. The utility of the new alternative can be calculated by summing together the known utilities for each of its specific attribute levels.

TABLE 8
Store Choice Probabilities

Store	Choice Probability
1	.04
2	.04
3	.04
4	.10
5	.01
6	.04
7	.08
8	.14
9	.10
10	.20
11	.04
12	.04
13	.04
14	.07
15	.07

The entire set of choice probabilities can then be recalculated for the enlarged store set. Similarly, if the attribute levels of an existing alternative were altered, a new utility and choice probability could be calculated for it. The easy accommodation of new alternatives or changes in existing ones is perhaps the greatest strength of conjoint analysis procedures.

CONCLUSION

This paper has discussed problems and their resolution in the application of the conjoint measurement model to real-world spatial choice, with special reference to choice behavior on recurrent urban travel. The principal problems encountered are: (1) the definition of the choice set, (2) the determination of choice-set attributes as stimuli for the experiment, (3) the complexity of data-gathering procedures, (4) the independence-of-attributes assumption, (5) the additivity assumption, and (6) computational and interpretive questions. Each of these problems is examined in turn, and a case study illustrates their resolution.

Although this paper has explored some of the principal problems of using a conjoint measurement model, obviously, other questions arise in its application to spatial choice and other choice behaviors. Particularly, questions of sampling and difficulties in administering questionnaires have been left aside. Also, there has been no discussion of the interpretation of group weights as utilities for each attribute level. In addition, there has been no treatment of theoretical and practical difficulties with computing attribute levels and destination utilities and converting the latter into choice probabilities. These are all questions that merit further research, but space precludes their inclusion here. Nonetheless, this paper seems to have fulfilled its main aim of demonstrating the potential usefulness of conjoint analysis in research on real-world spatial choice.

NOTES

1. This assumes that destination choices are separable from other choices (for example, choice of mode). The evidence for nonseparability is still sparse [2]. Also, trips are considered single purpose, since the modeling of multiple-purpose trips has not yet advanced far.

2. This ignores problems of the interpersonal comparison of utilities [3, p. 51] and whether it makes sense to talk of group utilities at all.

3. This assumption means that adjustments in the attributes of destinations do not cause changes in the attributes that individuals use for destination evaluation. It also means that new destinations should not have levels of attributes that lie outside the ranges used in analysis. This seems to imply that conjoint measurement will not handle the spatial choice effects of major innovations that affect destinations (for example, major innovations in retailing industries like the introduction of self-service or telephone service). The model will, however, handle the more common case of incremental adjustments in destination attributes.

4. Store 8 in Table 2 was known by only 42% of the respondents. It was less familiar to them than the other fourteen stores, though considerably more familiar than any other store not on the list in the table. It was retained for further analysis, although it was recognized that the utility calculated through conjoint measurement for the store would be biased to some unknown degree. The biasing effects of differential familiarity with alternatives needs further research.

5. This result does, however, suggest that some interaction among the attributes may be present, contrary to the assumption in the model. The low θ statistics in Table 4 nonetheless show that this interaction is insignificant. The effects of attribute interaction on the results of applying the model is the subject of further research.

REFERENCES

1. Atkinson, R. C., G. H. Bower, and E. J. Crothers, 1965. *An Introduction to Mathematical Learning Theory.* New York: John Wiley and Sons.

2. Ben-Akiva, M., 1975. "Structure of Passenger Demand Models," a paper presented at the 53rd Annual Meeting of the Transportation Research Board, January 1975.
3. Brummel, A. C., and E. J. Harman, 1974. "Behavioral Geography and Multidimensional Scaling." Hamilton, Ontario: Department of Geography, McMaster University, Discussion Paper #1.
4. Burnett, K. P., 1973. "The Dimensions of Alternatives in Spatial Choice Processes," *Geographical Analysis, 5*, 181-204.
5. Burnett, K. P., 1977. "Perceived Environmental Utility under the Influence of Alternative Transportation Systems: A Framework for Analysis," *Environment and Planning, 9*, 609-24.
6. Davidson, J. O., 1973. "Forecasting Traffic on STOL," *Operational Research Quarterly, 24*, 561-69.
7. Fiedler, J. A., 1972. *Condominium Design and Pricing.* Chicago: Association for Consumer Research.
8. Fingleton, B., 1976. "Alternative Approaches to Modeling Varied Spatial Behavior," *Geographical Analysis, 8*, 95-102.
9. Girt, J.L., 1976. "Some Extensions to Rushton's Spatial Preference Scaling Model," *Geographical Analysis, 2*, 137-56.
10. Golledge, R. G., and G. Rushton, 1972. *Multidimensional Scaling: Review and Geographical Applications.* Washington, D.C.: Association of American Geographers, Commission on College Geography, Technical Report #10.
11. Green, P. E., 1974. "On the Design of Choice Experiments Involving Multifactor Alternatives," *Journal of Consumer Research, 1*, 61-68.
12. Green, P. E., and V. R. Rao, 1971. "Conjoint Measurement for Quantifying Judgmental Data," *Journal of Marketing Research, 18*, 355-63.
13. Green, P. E., and Y. Wind, 1974. "New Techniques for Measuring Consumers' Judgements of Products and Services." Philadelphia: Wharton School Working Paper, University of Pennsylvania.
14. Johnson, R. M., 1973. "Pairwise Non-Metric Multidimensional Scaling," *Psychometrika, 38*, 11-18.
15. Johnson, R. M., 1973. "A Simple Method for Non-Metric Regression." Mimeographed report, Market Facts, Chicago.
16. Johnson, R. M., 1974. "Trade Off Analysis of Consumer Values," *Journal of Marketing Research, 11*, 121-27.
17. Kostyniuk, L., 1975. "A Behavioral Choice Model of Urban Shopping Activity." Ph.D. dissertation, Department of Engineering, State University of New York at Buffalo.
18. Krantz, D. H., 1964. "Conjoint Measurement: The Luce-Tukey Axiomatization and Some Extensions," *Journal of Mathematical Psychology, 11*, 248-77.
19. Krantz, D. H., and A. Tverksy, 1971. "Conjoint Measurement Analysis of Composition Rules in Psychology," *Psychological Review, 78*, 151-69.
20. Lieber, S. R., 1973. "A Comparison of Metric and Non-Metric Scaling Models in Consumer Research," a paper presented at the Symposium on Multidimensional Scaling at the Annual Meeting of the Association of American Geographers, Atlanta, Georgia, April 1973.
21. Louviere, J., E. Wilson, and J. Piccolo, 1977. "Psychological Modelling and Measurement in Travel Demand: A State-of-the-Art Review with Application." Resource Paper, Workshop I, Third International Conference on Behavioral Travel Modeling, Tanunda, Australia, April 1977.;
22. Luce, R. D., and J. W. Tukey, 1964. "Simultaneous Conjoint Measurement: A New

Type of Fundamental Measurement," *Journal of Mathematical Psychology*, 1, 1-27.

23. Rushton, G., 1969. "Analysis of Spatial Behavior by Revealed Space Preference," *Annals, Association of American Geographers*, 59, 391-400.

24. Rushton, G., 1969. "The Scaling of Locational Preferences." In K. R. Cox and R. G. Golledge (eds.), *Behavioral Problems in Geography: A Symposium*. Evanston, Ill.: Department of Geography, Northwestern University, Studies in Geography #17.

25. Rushton, G., 1974. "Decomposition of Space Preference Functions," a paper presented at the Symposium on Multidimensional Scaling at the Annual Meeting of the Association of American Geographers, Atlanta, Georgia, April 1973. (Revised December 1974.)

26. Stopher, P., 1975. "Development of the Second Survey-Shopping Destination Study." Unpublished manuscript, Northwestern University, Evanston, Ill.

27. Stutz, F., 1975. "Environmental Trade-offs for Travel Behavior," a paper presented at the Annual Meeting of the Association of American Geographers, Milwaukee, April 1975.

28. Tversky, A., 1967. "A General Theory of Polynomial Conjoint Measurement," *Journal of Mathematical Psychology*, 4, 1-20.

29. White, R. W., 1976. "A Generalization of the Utility Theory Approach to the Problem of Spatial Interaction," *Geographical Analysis*, 8, 39-46.

30. Wind, Y., S. Jolly, and A. O'Conner, 1975. "Concept Testing as Input to Strategic Market Simulations," a paper presented at the 58th International Conference of the American Marketing Association, Chicago, April 1975.

31. Young, F. W., 1972. "A Model for Polynomial Conjoint Analysis." In A. K. Romney, R. N. Shepard, and S. B. Nerlove (eds.), *Multidimensional Scaling: Theory and Applications in the Behavioral Sciences* (vol. 1), pp. 69-104. New York: Seminar Press.

Chapter 2.5 Applications of Functional Measurement to Problems in Spatial Decision Making

Jordan J. Louviere

Many problems of interest to geographers, planners, and other spatial scientists can be viewed as problems in human decision making, judgment, evaluation, or choice. Many such judgments are of direct interest in that they involve a judgment or decision to move, to travel, to locate, et cetera. Others are of indirect interest in that they are not concerned directly with spatial behavior but may have consequences for spatial behavior. An example of the latter is a planner's evaluation of zoning requests for land-use classification changes by a planning commission.

This paper deals with a particular theory of how individuals make decisions, evaluations, judgments, et cetera, and a methodology by which the theory can be applied to problems of geographical interest. As a vehicle for facilitating the discussion, a paradigm for the study of spatial behavior in general is outlined. This paradigm is then employed to explicate the role that functional measurement can play in advancing our understanding of spatial behavior. Examples of applications are given from psychophysics, transit planning, and spatial behavior.

I would like to thank Larry Ostresh, Bob Meyer, and Mike Piccolo for reading earlier drafts of this paper, editing, and providing useful commentary. Particular thanks are due to Mike Piccolo for assistance in data collection and analysis and for seeing the final version through typing and mailing in my absence. Any mistakes that remain are solely my responsibility.

THEORY

Functional measurement, or stimulus integration, theory was developed by Norman Anderson during the early 1960s. It has rapidly acquired adherents and has built up a considerable body of theoretical and empirical research, much of which is summarized in a series of recent publications by Anderson [1-8]. The theory is concerned with how individuals put together or combine (integrate) several different pieces of information (stimuli) into an overall judgment, evaluation, et cetera, about some phenomenon under investigation. The functional measurement paradigm holds that the separate stimuli are combined "as if" the individual employed an algebraic rule or function to integrate them. Fundamentally, therefore, functional measurement is concerned with the behavior of algebraic models, and this makes it useful for studying behavioral phenomena of many kinds.

As Anderson and others have argued [1-8, 17], the development of empirical laws relating the behavior of a set of independent variables to that of a dependent variable should be the primary focus of a science of psychology, rather than measurement per se. Once an empirical law is derived, the law provides the basis for psychological measurement or scaling, rather than the latter providing the basis for the former. Thus, the label "functional measurement" refers to the fact that, once an empirical law is known or assumed between stimuli and response, this knowledge or assumption permits one to derive measurement scales for the stimuli on the response dimension. These measures, in essence, are those that "function" in the particular empirical law in question.

A PARADIGM FOR RESEARCH INTO
SPATIAL BEHAVIOR

These concerns will now be stated in a more formal manner by introducing a general view of the entire stimuli → response → behavior system. This view is a comprehensive one and is not intended to be limited to problems of a geographic nature, although constant reference will be made to such problems. I will present it as a scheme adapted from the separate works of Anderson and Birnbaum [7, 8, 9] (Figure 1).

The first box is defined as the Ψ box; it is concerned with the relationships among physical values or measures of influential or

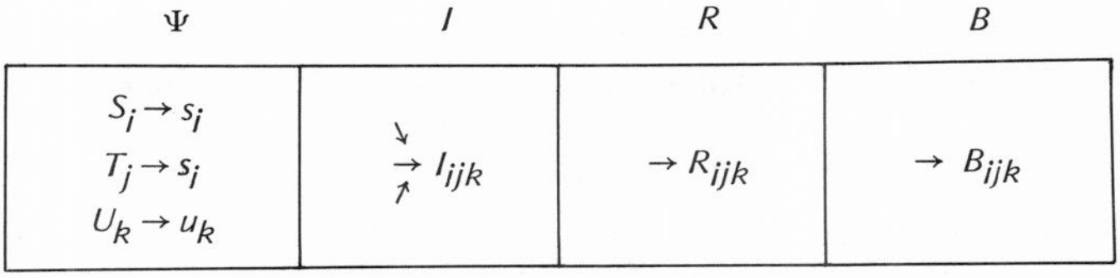

Ψ	I	R	B
$S_i \to s_i$ $T_j \to s_j$ $U_k \to u_k$	$\searrow$ $\nearrow$ I_{ijk}	$\to R_{ijk}$	$\to B_{ijk}$

Figure 1. A paradigm of spatial behavior

causal variables (if such quantitative measures exist) and corresponding subjective or psychological values of the same. Thus:

$$s_i = \Psi_1(S_i)$$
$$t_j = \Psi_2(T_j) \qquad (1)$$
$$u_k = \Psi_3(U_k)$$

where uppercase letters refer to physical values such as actual size of town, actual length of a line, base of a triangle, et cetera, and lowercase letters refer to corresponding psychological values such as perceived base of triangle, et cetera. Although considerable work in psychology has been directed at these relationships, little work other than that of Burnett, Cadwallader, Hensher, and Louviere [10, 11, 15, 22] is evident in geography and related areas. These researchers have consistently argued for power function relationships; I will present empirical data later in this paper in support of same.

The second box is the integration box. It deals with the rule or model by which the psychological values are combined. Because combination rules are important in science, this box assumes the role of a model or theory by which the subjective values are "integrated" or combined into a single evaluation, judgment, or impression of the phenomena as a whole. it is the integration of influential stimuli, therefore, that permits one to state verbally, "I like this store so many units better than that one" or "that weight is heavier than this one," et cetera. This is an "as if" process that involves finding the empirical rule that relates the overall impression to the separate stimuli. It has been the object of considerable research in psychology, and excellent reviews are available (Anderson [3, 4, 6] and Slovic and Lichtenstein [33]. Recent work in geography by Louviere [22, 25], Schmitt [31], Lieber [20], Stutz [34], and Knight and Menchik [16] has focused on this problem; but the results do not permit any general conclusions regarding combination rules at this time, although a multiplicative rule seems to be favored. It should be noted that the subjective or psychological combination value, I_{ijk}, cannot be observed; nonetheless, I can still make assumptions about it.

The third box is the response box. It is here that some overt response is observed. This is usually some number given on a psychological response dimension—say, "willingness to shop at a certain place" or "perceived length of line." I will put aside for the moment the issue of the level of such measurement and assume that some function exists to map I_{ijk} into R_{ijk}. I would normally like to infer the relationship between the physical stimuli and their psychological values as well as the combination rule that obtains by analyzing the response data.

The last box is the behavior box. This box postulates a relationship between the psychological response measure and overt behavior. The existence of a transformation relation, of course, would permit investigators to make statements about human behavior from laboratory experiments or field data. This would permit the development of causal models that simulate the behavior of the system in situations in which it is practically impossible to experiment—the situation in much of human geography.

Any number of assumptions can be made about the Ψ, I, R, and B functions. To begin with, I will make some simple assumptions that are widely used in psychological modeling and derive their consequences. In particular, I assume the following general forms:

$$\Psi = x_i = a_i + b_i X_i^{c_i} \quad (i = 1, 2, \ldots n) \tag{2}$$
$$I = I = \sum_i (x_i) \text{ or } \prod_i (x_i) \tag{3}$$
$$R = R = a_{n+1} + b_{n+1} I \tag{4}$$
$$B = B = a_{n+2} + b_{n+2} R + b_{n+3} R^2 \tag{5}$$

where X_i stands for any of the n experimental variables or stimuli; x_i stands for the psychological values of X_i; and the a's, b's, and c's are constants to be derived or determined empirically. I have written the general form of a power function with an intercept for Ψ, I, R, and B and have assumed a linear or a multiplicative combination rule for I. Many others are possible, but these represent common assumptions that permit one to derive some useful consequences.

In particular, let us treat the two-variable cases in which there are two physical variables, S_i and T_j. These might represent, for example, town size and distance. One common set of assumptions might be:

$$\Psi(s_i) = a_1 + b_1 S_i^{c_1} \tag{6}$$
$$\Psi(t_j) = a_2 + b_2 T_j^{c_2} \tag{7}$$
$$I_{ij} = s_i + t_j \tag{8}$$
$$R_{ij} = a_3 + b_3 I_{ij} \tag{9}$$
$$B_{ij} = a_4 + b_4 R_{ij} + b_5 R_{ij}^2 \tag{10}$$

By substitution:

$$B_{ij} = a_4 + b_4 \left\{ a_3 + b_3 [(a_1 + b_1 S_1^{c_1}) + (a_2 + b_2 T_j^{c_2})] \right\} \tag{11}$$
$$+ b_5 \left\{ a_3 + b_3 [(a_1 + b_1 S_i^{c_1}) + (a_2 + b_2 T_j^{c_2})]^2 \right\}$$

which represents the most general form of the derivation. In actuality, there would be fewer parameters than the eleven in equation 11, because once equation 11 is expanded and simplified, the following can be derived:

$$B_{ij} = k_0 + k_1 S_i^{c_1} + k_2 S_i^{2c_1} + k_3 T_j^{c_2} + k_4 T_j^{2c_2} + k_5 S_i^{c_1} T_j^{c_2} \tag{12}$$

Note that, if equation 10 is any Taylor Series expansion beyond degree one, the data should appear to be multiplicative, even if it were generated by an additive rule in equation 8. This, of course, is a most likely case, so it provides a potential argument for multiplicative-looking data in the response and behavior metrics. Moreover, it also suggests that one can approximate this overall equation 12 by means of a generalized Taylor series. This further implies that commonly applied multivariate methods, such as multiple linear regression, can be employed.

The most obvious case, of course, is to let the combination rule in equation 8 be multiplicative, in which case the data will not only look multiplicative, they will *be* multiplicative. This does not constitute a case for multiplication, so it is a trivial case unless it can be deduced on other grounds. One way to do this is to consider the process itself. The next sections consider various judgment processes and their applicability to areas of interest to psychologists and geographers. It will be argued that functional measurement is one of several tools that can be useful in diagnosing the relationships implied in Figure 1, as well as in testing hypotheses generated from a priori considerations.

Linear Processes for Integration Functions

I will restrict my attention to adding and averaging models as linear processes, although there are other possibilities. In the discussion of models and derivations therefrom to follow in this and later sections, I will continue to use the notation developed in relation to Figure 1. Because it will frequently be the case that there are no physical referents (the Ψ box), I will be treating the case of equations 3 and 4; that is, the general relationship between the integration of psychological values and responses will be the focus of interest.The transformation from physical values to responses should be understood to be of secondary interest in the examples provided. The general linear judgment model for three independent variables and a dependent response measure can be written as follows, where all variables are measured in units of R:

$$R_{ijk} = s_i + t_j + u_k + \epsilon_{ijk} \tag{13}$$

where R_{ijk} is a dependent response observed on some numerical scale and s_i, t_j, and u_k are independent variables measured in the subjective response metric. The subjective values can be related in a variety of ways to corresponding physical measures. They also can be totally qualitative and have no physical referent. Thus, equation 13 represents the general form for qualitative or quantitative independent variables measured subjectively and is one possible combination of equations 8 and 9.

If the R_{ijk} have been observed as the outcomes of a factorial experiment, I can prove that the marginal means of the response data constitute interval-level measures of the psychological values in equation 13. Averaging equation 13 over columns and surfaces, I obtain:

$$R_{i..} = s_i + \bar{t} + \bar{u} + \epsilon_{i..} \tag{14}$$

Thus, s_i is equal to the marginal mean of the rows up to a linear transformation. Corresponding logic holds for t_j and u_k. Hence, one set of values for s_i, t_j, and u_k is given by the marginal response means.

Equation 13 can be tested by subjecting the data to an analysis of variance because the equation implies that there will be nonsignificant two-way and three-way interactions. This constitutes both a rationale for diagnosis and a statistical test for an hypothesized model. If the independent variables have physical referents, estimates of the Ψ function can be obtained by plotting the marginal response means for rows, columns, or surfaces against their corresponding physical counterparts. No a priori knowledge or measurement is necessary to establish these relationships; those relationships are derived that "function" in the establishment of the empirical law.

When one has a set of physically measured variables—say, S_i, T_j, and U_k—the logic of equations 2 to 4 can be applied, to derive the following expectation:

$$s_i = a_1 + b_1 S_i^{c_1} \tag{15}$$
$$t_j = a_2 + b_2 T_j^{c_2} \tag{16}$$
$$U_k = a_3 + b_3 U_k^{c_3} \tag{17}$$
$$\Psi_{ijk} = s_i + t_j + u_k + \epsilon_{ijk} \tag{18}$$
$$R_{ijk} = a_0 + b_0 \Psi_{ijk} \tag{19}$$
$$R_{ijk} = [(a_1 + b_1 S_i^{c_1}) + (a_2 + b_2 T_j^{c_2}) + (a_3 + b_3 U_k^{c_3})] b_0 + a_0 \tag{20}$$
$$= K_1 + K_2 S_i^{k_3} + K_4 T_j^{k_5} + K_6 U_k^{k_7} \tag{21}$$

Thus, strictly, linear regression is inadequate to account for response behavior even when one assumes an additive integration function (equation 18) unless $K_3 = K_5 = K_7 = 1.0$. This logic holds for adding, averaging, or subtracting models.

In the psychological response metric, it is impossible to distinguish between adding and averaging. Averaging is a special case of addition in which the sum of the weights equals unity. Integration theory assumes both weight and scale value parameters. Hence, the general linear form can be written as:

$$R_{ijk} = w_1 s_i + w_2 t_j + w_3 u_k + \epsilon_{ijk} \tag{22}$$

But the w's cannot be uniquely estimated without a priori knowledge or constraints, because the s_i, t_j, and u_k are generally unknowns. The case in which their counterparts S_i, T_j, and U_k are known and measured without error and c_i in equation 2 equals unity is a classic linear regression situation. In general, this will not be the case; hence, it might be preferable to seek subjective values. Thus, in the case of deriving psychological measures of the independent variables, one might seek least-squares estimates of the psychological parameters. But this is equivalent to least-squares estimates of the "effects" of the levels of the independent variables.

These deductions from equations 15 to 21 have a geometric counterpart. The absence of interactions translates into the expectation that each n-way plot should consist of a series of parallel lines. To see why this should be the case, consider the result of subtracting row 1 from row 2 in equation 14 or equation 18.

$$R_{1jk} - R_{2jk} = (s_1 + t_j + u_k) - (s_2 + t_j + u_k) \tag{23}$$
$$= s_1 - s_2 = k \tag{24}$$

Similar logic holds for equation 21. Thus, the result of subtracting any two rows from one another is a constant difference. This is equivalent, of course, to saying that all lines have the same slopes but different intercepts. Similar logic holds for columns and surfaces.

Empirical Applications

I will provide one psychophysical application as a general discussion of linear processes, two geographical examples, and a fourth example from person perception in social psychology.

Ten student subjects participated in an experiment designed to show that functional measurement can recover a hypothesized additive relationship. Subjects were shown combinations of two line lengths drawn on half of a sheet of typewriter paper and arranged in a 3 x 7 factorial design (rows: 3 centimeters, 5 centimeters, 7 centimeters; 2 centimeters, 4 centimeters, 6 centimeters, 8 centimeters, 10 centimeters, 12 centimeters, 14 centimeters). The position (left, right) of the row and column stimulus was randomized, as was the order of presentation of the twenty-one combinations.

Subjects were shown a standard stimulus that consisted of two lines 1 centimeter long. Subjects were told to call that combination a "2." They were then free to use any number they wanted to judge the "combined length" of the left and right stimulus lines. Subjects were told to mentally place one line on top of the other and judge "how long" they thought the combined length would be relative to the standard stimulus, which was displayed continually throughout the experiment. Subjects judged each set of lines twice in a within- and between-subjects design.

Data were analyzed initially by an analysis of variance, and they were graphically plotted (Figure 2). The interaction term for rows x columns did not approach significance ($p > .10$), and the plotted data clearly exhibit parallelism. Thus, it is safe to conclude that the subjects acted "as if" they added the row and column stimulus to judge combined length. Some may argue that this is an obvious result. The result, however, could only be obvious if the subjects were employing a linear strategy and the response scale were at least a valid interval scale. Thus, the results simultaneously verify the response scale and the model. Data were then fit by multiple linear regression procedures for the following two models:

$$R_{ij} = k_1 + k_2 s_i + k_3 t_j + \epsilon_{ij} \tag{25}$$
$$R_{ij} = k_4 + k_5 S_i + k_6 T_j + \epsilon'_{ij} \tag{26}$$

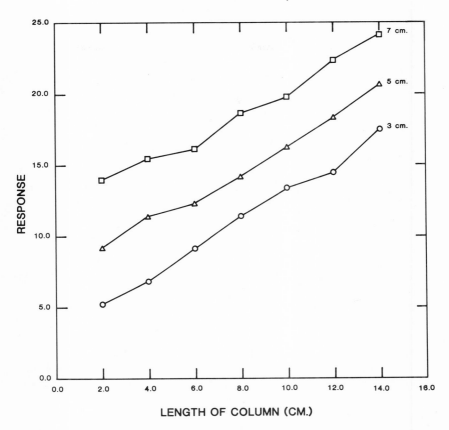

Figure 2. Combining time lengths: analysis of variance results

Equation 25 becomes:

$$R_{ij} = s_i + t_j - 14.8135 \quad (R^2 = .994) \tag{27}$$

where s_i and t_j are the marginal means. That is, the values k_2 and k_3 are unity and k_1 is 14.8135. The grand mean is 14.8119, so the program recovers the relationship:

$$R_{ij} = R_{i.} + R_{.j} - R_{..} \tag{28}$$

which can be derived from equations 13 and 14 by substitution. Thus, the analysis further confirms that interval nature of the response scale. Equation 26 becomes:

$$R_{ij} = 1.8732 S_i + 0.92887 T_j - 1.98512 \quad (R^2 = .990) \quad (29)$$

which is the direct estimation of R_{ij} from the physical metric.

Had I not had a hypothesized model a priori, I could have inferred a linear process from the data analysis. Thus, the power of the functional measurement approach to diagnose algebraic combination rules is obvious. If psychological measurement were a goal, it could be accomplished for the column values by letting the marginal mean for 2 centimeters equal 0, and that for 4 centimeters equal unity. Then, one estimates the linear relationship between them: intercept equals marginal mean of 2 centimeters; slope (or unit of scale) equals $R_{.4cm} - R_{.2cm}$. The remainder of the values on the interval scale can then be recovered. Another method would be to use the following transformations:

$$R_{i.} = a_1 + b_1 S_i \quad (30)$$
$$R_{.j} = a_2 + b_2 T_j$$

Louviere and Meyer [24] studied student judgments of the residential desirability of neighborhoods. From a large number of photographs of houses, Louviere and Meyer selected sets of three that were approximately equal in subjective value according to a prescaling task (three high values; three moderate values; three low values). A 3 x 3 x 3 factorial design was constructed in which each row, column, and surface consisted of a high-, moderate-, and low-value photograph. Subjects judged all twenty-seven sets of three photographs at a time on a scale from 1 to 20 in which 1 was the worst possible "neighborhood" imaginable and 20 was the best. Subjects were told that each set of three photographs represented a sample of three houses (e.g., HLH, MHH) from a neighborhood in some city in southern Florida.

Based on previous evidence compiled by Anderson in person perception [6, 7, 8], it was hypothesized that the subjects would act "as if" they averaged the three photographs to make a response to the set of three. An analysis of variance revealed only one significant interaction for one subject. This interaction was traceable to a reversal of photo valued for this one subject: the low photo on one dimension was judged to be higher in value than the moderate. When the photos were reversed, the interaction disappeared. Thus, for all subjects the hypothesis of linear averaging was retained. Plots of the data also revealed systematic parallelism confirming the results of the statistical tests (Figure 3).

Stutz [34] studied the manner in which students evaluated

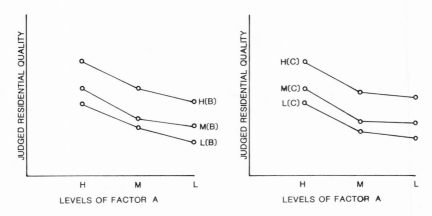

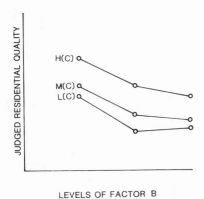

LEVELS OF FACTOR B

Figure 3. Results of linear averaging experiment on judged residential quality.
(The first graph shows the average ratings given to combinations of house photos
in factors A and B, with the photos in C held constant. The second graph shows
the average ratings of house photos in factors A and C, with the photos in B held
constant. The third graph shows the average ratings given to combinations of
house photos in factors B and C, with the photos in A held constant. The graphs
illustrate the outcome of an experiment in which subjects judged combinations
of photographs of houses. Each combination of three houses was constructed from
a 3 x 3 x 3 factorial design in which each of the three factors had three levels.
The levels were pictures of houses preselected on the basis of ratings, which
classified the houses as high, medium, or low in residential quality.)

potential beaches were constructed from three levels each of crowding, distance, and cleanliness to comprise a 3 x 3 x 3 factorial experiment. Subjects judged the probability that they would choose each hypothetical beach on a 150-millimeter line scale. Subjects' responses were measured to the nearest millimeter and analyzed by means of an analysis of variance. As in the Louviere and Meyer example, all two- and three-way interactions were nonsignificant, permitting Stutz to retain the linear processing hypothesis.

Anderson and his students [3, 4, 6, 7, 8] have consistently found averaging processes in studies of person perception. Particularly in experiments in which subjects are asked to judge the likeableness of persons who are described by combinations of personality-trait adjectives (e.g., bold, easy-going, humorless), linear processing has consistently emerged. Follow-up studies have consistently provided support for averaging over adding formulations. The test is qualitative and depends *only* upon ordinal assumptions. When two high-valued adjectives are judged by themselves and then in combination with two medium-valued adjectives (HH, HHMM), an adding model must predict an increase in the rank of the response because of the presence of additional information that should, in theory, be added. The averaging hypothesis, in contrast, predicts a lowering of the response toward the medium-valued adjectives. Results consistent with averaging have consistently been obtained, with no support for addition.

It should be noted that this result has important implications for rank-order methods such as conjoint measurement, since they are unable to distinguish averaging from general linear processing. Even though such models have only "as if" status, it is important to be able to distinguish between various "as if" processes, and it is precisely at this point that the ordinal methods falter because they are insensitive to such "differences that make the difference." Later I will take up a particularly troublesome case for the ordinal procedures—that of distinguishing multiplication from differential weighting in an averaging context. The power of metric approaches such as functional measurement will then become evident.

Differentially Weighted Averaging Processes

I will provide a discussion, but no examples, of the application of this model because no current geographical examples exist. The differentially weighted averaging model is a configural model in that it permits the weight values to vary as a function of the stimulus values. In this way, nonlinear effects can be obtained. The model can be written as follows for three independent variables:

$$R_{ijk} = K + \frac{w_i s_i + w_j t_j + w_k u_k}{w_i + w_j + w_k} + \epsilon_{ijk} \tag{31}$$

where k is a scale constant to allow for an arbitrary zero; the w's are weights attached to the s_i, t_j, and u_k stimulus dimensions, respectively; and ϵ_{ijk} is an additive error term.

Note that the special case in which the weights within each dimension are constant (i.e., all w_i's are equal; all w_j's; all w_k's) produces the constant weight or linear averaging model. For values of w_i, w_j, and w_k that are not constant within their respective dimensions, there is a high probability of producing nonlinearities, including crossover interactions. If, for example, the w_i's had high values for the two extreme values of s_i and moderate or low values for moderate values of s_i and w_j's are moderate to low for extreme t_j and high for moderate t_j, crossover interactions would be expected in the interaction "effects" of these two factors. These crossovers can be handled by the differentially weighted averaging model but are troublesome "errors" in the ordinal or conjoint measurement approach.

The model says that the individual takes a weighted average in each response trial, but each weighted average is different because $w_i + w_j + w_k$ is different in each cell. To the extent that the w's are well behaved with respect to the corresponding stimuli, the model is not only estimable but useful in a predictive sense. Although no examples of applications of this model have arisen in geographic or related work, Anderson [6, 7] has consistently found this model to obtain in impression formation work. Typically, for judgments of propensity to date another person when the subject is presented with a picture of a person accompanied by personality-trait adjectives, extreme values of the photograph are given heavier weight and the adjectives are given little or no weight, except with moderate- to low-valued photographs. Frequently, the pattern of weighting, when heavier toward the undesirable end of the scale, can produce data that look multiplicative. Once again, the ordinal methods incorrectly diagnose the data as multiplicative when they are not.

Shanteau [32] has shown, as in the adding-versus-averaging case, that a simple directional test can distinguish between the two rules.

Employing a Bayesian inference task, Shanteau argues that the critical test revolves around the presentation of neutral information. Given two probability urns (thirty/seventy white/red and seventy/thirty white/red) and a sample stimulus of one white chip, the subject judges the probability that it came from the white (seventy/thirty) urn. Subjects typically respond around 60%. Then, presented with a sample of three red and three white chips that is logically neutral in information, the subject judges the probability of it coming from the white seventy/thirty urn again. This sample has a likelihood ratio of unity and, hence, when combined with the previous sample should leave the response unchanged at 60%. An averaging process, on the other hand, predicts a downward change in the response, because the new information has a value of only 50%. In three separate experiments in which subjects responded on a probability scale and a log-odds scale, the shift in the predicted direction was observed for almost all subjects in every condition. Thus, the functional measurement approach can distinguish between these decision rules and can model them effectively with algebraic procedures.

Multiplying Processes

Multiplying processes are of theoretical interest to geographers and related researchers because a number of such forms have been proposed to represent spatial behavior. One obvious example is the gravity model of social physics. Let us consider an experiment in which subjects judge willingness to shop in towns (outside of their own) that vary in selection (number of shops) and travel time away (hours, minutes). If subjects respond on some continuous numerical scale, one can write:

$$R_{ij} = k + s_i t_j + \epsilon_{ij} \tag{32}$$

where R_{ij} is the observed response to combinations of selection and travel time; k is an additive constant allowing for an arbitrary zero in the response scale; s_i and t_j are, respectively, subjective values of selection and travel time; and ϵ_{ij} is an additive error term (see Anderson [3, 4, 6] for a justification of the additive form).

This model can be tested by an analysis of variance when the levels of selection and travel time are varied in a factorial design. This model predicts a significant interaction between selection and travel time. In particular, it predicts that all of the variance in the interaction term will be concentrated in the linear x linear component of the interaction. This amounts to predicting that only the

bilinear term will be significant and the residual will be nonsignificant.

A geometric test is also available. Consider the result of subtracting row one from row two:

$$R_{1j} - R_{2j} = (k + s_1 t_j + \epsilon_{1j}) - (k + s_2 t_j + \epsilon_{2j}) \qquad (33)$$

$$= t_j(s_1 - s_2) + (\epsilon_{1j} - \epsilon_{2j}) \qquad (34)$$

When the graph is spaced so that the lowest-valued element of t_j is
on the right, equation 34 predicts that the difference between each
row will not be constant but will differ at each point s_i according to
the value of t_j. In this instance, the curves converge to the lowest
value of t_j. Thus, in contrast to parallelism, the model predicts a
diverging fan of curves. The curves should be straight lines when
the graph is plotted in the response metric.

As in the linear case, the marginal means of the response data
can be shown to constitute one set of measures for s_i and t_j. Consider the result of averaging equation 32 over columns:

$$R_{i.} = k + s_i \bar{t}_. + \epsilon_{i.} \qquad (35)$$

And s_i is equal to $R_{i.}$ up to a linear transformation. Hence, $R_{i.}$ must
be at least on an interval scale.

Empirical Applications

A number of applications of the multiplying model of geographic interest are available. I will consider the work of Norman and
Louviere [28], Lieber [20], and Piccolo and Louviere[29].

Norman and Louviere [28] examined the effects of varying fare,
frequency of service, and walking distance on judgments of the
subjective likelihood of taking a bus in Iowa City, Iowa. Twenty
student subjects were presented with twenty-seven combinations
of fare, service, and walking distance in a 3 x 3 x 3 factorial design.
Each combination described a different hypothetical bus system.
Subjects judged each bus system description on a 150-millimeter
line scale; responses were scored to the nearest millimeter.

Data from the experiment were graphically plotted (Figure 4)
and analyzed by means of an analysis of variance. All the two- and
three-way interaction terms were significant ($p < .01$) and concentrated in the linear-by-linear components of the respective interactions. A multiplying model of the form:

$$R_{ijk} = f_i \cdot s_j \cdot w_k \qquad (36)$$

was fit to the data by the method of maximum likelihood for the case of normally distributed parameters with unequal variances using a computer program developed by Normal [27]. This model fit considerably better than a simple linear or general polynomial regression when each is compared in terms of reduced sums of squares. Thus, it was concluded that subjects made probability decisions regarding using the bus "as if" they multiplied the subjective values of fare, service, and walking distance. In addition, it was shown that the Ψ function for each variable was nonlinear. With only three points, however, it is difficult to say more about the respective functions.

Lieber [20] studied the effects of varying three levels each of distance to close friends and relatives, distance to fresh-air

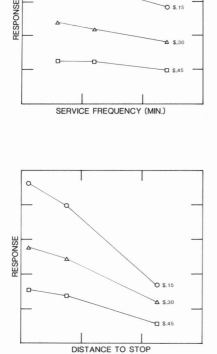

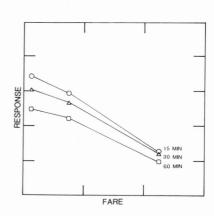

Figure 4. Iowa City bus experiment.

recreational opportunities, and distance to major urban areas on judgments of desirability of places as migration destinations for a first job. Subjects were graduating college seniors who were asked to assume that they had a number of job offers—comparable in salary—that differed in values of the three variables. Subjects made judgments about how desirable each alternative was compared to all others on a 150-millimeter line scale.

Data were analyzed via an analysis of variance for factorial design and were also rank ordered and analyzed with Young's POLYCON computer program for conjoint analysis. Both analyses agreed almost all of the time that the subjects acted "as if" they multiplied the subjective values to make a desirability response. Plots of the data were in accord with the geometric expectation that the data would look like a diverging series of lines. Subjective values of distance did not appear to be linear functions of experimental distance values.

Piccolo and Louviere [29] examined two gravity-model-type decision processes: (a) evaluations of towns and cities other than one's own as places to shop and (b) evaluations of towns and cities as places to live, given that one has taken a job at a rural manufacturing plant or mining establishment and cannot live "on-site." In both of these cases, the relationships estimated in the laboratory were correlated with observed data in the field with an entirely different subject population. In both instances, the laboratory-derived relationships correlated well with data on real choices in the field.

Piccolo and Louviere [29] presented 110 undergraduate students at the University of Wyoming with twenty-five combinations of travel time by automobile and number of stores (five levels each: ½, 1, 2, 3, 4 hours; 10, 50, 100, 200, 400 stores), each of which described a different town. Subjects were asked to evaluate each of these towns relative to Laramie, Wyoming, as a place to shop, on a 100-point scale. All subjects completed all twenty-five evaluations. Analysis of the data revealed a highly significant travel-time-by-number-of-stores interaction, and the graph of this interaction clearly indicated a multiplicative form, as deduced in equation 34. A simple multiplicative expression of the general form of the equation 32 fit the cell means very well.

A random sample of the 250 individuals in Laramie, Wyoming, was drawn from the local telephone directory, and these individuals were asked whether or not they had shopped in each of eleven places outside of Laramie during the past week. Proportions of the sample's choices of the eleven places became the dependent variable or measure of "real" behavior. Travel times from Laramie to each town and the number of retail stores were estimated for each town from maps and census data. The number of stores and the travel time were substituted in the laboratory-derived equation and used to predict the expected numerical evaluation that the sample would have given that combination had it been present in the experimental design. These expected numerical evaluations were graphed against the corresponding choice proportions, and the rank-order correlation was estimated. The correlation was .73, and the graph revealed a scatter of points of a generally positively accelerated form. Inspection of residuals from a regression analysis suggested the omission of an important spatial variable: the location of each town with respect to one or more others on the same route, or intervening opportunities. This was measured as the cumulative number of stores that would have to be bypassed by the individual to shop at a particular destination following the most direct route. After controlling for expected evaluation in a multiple regression, the intervening opportunities measure was found to be highly statistically significant, confirming the aforementioned observation about the residuals and suggesting that it must be included in future laboratory studies. These data, of course, suggest that there is a relationship between the laboratory-derived decision model and real behavior. This relationship was pursued further in the next study.

Piccolo and Louviere [29] presented seventy-five University of Wyoming faculty members, staff members, and students with thirty-six combinations of commuting distance in miles and population plus a description of shops, services, and facilities available (six levels each: 15, 30, 45, 60, 75, 90 miles; and 200, 500, 1000, 1500, 2000, 2500 people plus amenities). Subjects were randomly assigned to five different conditions in which a different random order of presentation of the thirty-six combinations was used to test for order effects. Subjects were asked to place themselves in the role of taking a job at a mine or a plant in rural Wyoming, where they could not live but must select some town or rural area as a place to live. As in the shopping study, they were asked to express their degree of preference for each of the thirty-six towns, on a 100-point scale.

The results indicated that there were no order effects and, as before, revealed a highly significant distance-by-town-size interaction. Graphs of the interaction were again consistent with multiplication, and a simple multiplicative expression was derived that reproduced the cell means very well. Data were available for a number of plants and mines in the Rocky Mountain West in which employees indicated their current town of residence, and the location (in miles) of the work place from the town was recorded. Estimates of town populations were obtained from the latest available state and federal sources (it is difficult to say how accurate available estimates are because of the very rapid growth rates of most of these "boom" towns). Commuting-distance and town-size estimates were substituted in the equation derived from the laboratory study to derive expected numerical preference estimates for each real town ($n = 56$).

As in the shopping study, the predicted numerical preference values for each town were graphed against the proportion of non-local employees at each plant or mine who chose each town that was available, and the rank-order correlation between the measures was computed. The graph was similar in form to the shopping graph, and the relationship was much stronger ($r = .93$). No relationship with intervening opportunities was found, despite a number of different tests for its presence. In this instance, the laboratory-derived preference values correlate very well with the data, but the relationship is clearly not linear. Follow-up work with these data by Lerman and Louviere [18] have established that the functional form uncovered in the laboratory outperforms a very large number of potential competitors that would usually be estimated in econometric applications of logit analysis, treating the data as a discrete choice problem in which repeated observations on the same alternatives are available. The best-fitting logit model using the laboratory-derived functional specification in fact does no better in terms of reproducing the data than simply fitting the predicted preference values to the data. Hence, this study indicates very good correspondence between laboratory observations and real-world observations.

Recent Tests of Correspondence of Laboratory Models with Behavior

A number of recent studies have compiled further evidence of the correspondence of laboratory-simulated and real behavior. In particular, two studies by Meyer, Levin, and Louviere [26] have both demonstrated that the use of public transit in Iowa City, Iowa, by University of Iowa employees is highly related to their evaluations of transit systems that differ in terms of attributes relevant to choice of transportation mode. Louviere and Meyer [5] have shown that choice of grocery stores in Laramie, Wyoming, Tallahassee, Florida, and Iowa City, Iowa, is highly related to a simple multiplicative combination of separate evaluations of price, selection, and convenience of each store by consumers. Further, at least for price and selection, there are very strong relationships between aggregate evaluations and physical measures of these variables at each store, demonstrating the link to objectively observable phenomena. More recently, Louviere [23] has demonstrated that models describing consumer evaluations of transit systems in Xenia, Ohio, very accurately predict the actual patronage of the Xenia public transit system since its inception in 1974. This latter study is a particularly good test of the approach, because the Xenia transit system underwent rather significant changes from year to year, including shifts from buses to jitneys to paratransit as part of a U.S. Department of Transportation model transit demonstration program. Currently, research is underway at the University of Iowa to further extend the approach to predicting transportation mode choices in a rapidly changing energy environment and to predicting the retirement migration choices of the fifty-to-sixty-five-year-old age-group. Other work involves the estimation of the demand for international and domestic air travel as a function of the availability of new budget air fares to various destinations and the estimation of the demand for new mortgage instruments in the housing market. Although not a complete enumeration of applications already undertaken or underway, the preceding discussion should provide a reasonable overview of the method, its current empirical status, and its future potential.

DISCUSSION AND CONCLUSIONS

This paper has discussed the theories and methods employed in the study of human information processing in making judgments. In particular, attention was focused on the stimulus integration theory of functional measurement as one vehicle by which to study decision processes. Functional measurement differs from ordinal approaches to the development of mathematical descriptions of judgments in that it relies on numerical judgments, analysis of variance, and algebraic models as descriptors of process. However, as Anderson has repeatedly pointed out [1-8], functional measurement is not restricted to numerical responses. It allows for monotonic rescaling of ordinal responses to satisfy a hypothesized algebraic model. Yet, it has rarely been needed. The experimental procedures developed to implement the functional measurement approach, which were not discussed in detail in this paper, have proved to be of considerable value in satisfying the numerical assumptions, as evidenced by the record of consistent empirical success reported by Anderson and his associates.

As more and more empirical results demonstrating correspondence with real spatial behavior are amassed, this approach should become attractive to researchers interested in spatial behavior. Evidence of correspondence was supplied in the body of this paper; however, I suggest that further evidence is necessary to establish generality outside the laboratory. Given such evidence, decision models derived via functional measurement can serve as axioms in spatial theories. For example, if one had a decision model that described how consumers make decisions about where to shop in a specific context, say, groceries, one could use this, together with sets of initial conditions (spatial configurations) and axioms about entrepreneurial behavior, to deduce the state of spatial systems.

In a more fundamental sense, these procedures could be very useful in deriving decision rules that do not depend upon the particular conditions of a unique spatial system. In particular, the argument of Curry [12] and Rushton [30] regarding the uniqueness of parameters estimated on specific spatial systems could potentially be satisfied. Both have argued that we need to derive models that predict spatial behavior regardless of the configuration of the independent variables in the system. Although Rushton [30] has proposed an approach to do this via multidimensional scaling, MDS, too suffers from the fact that no information is available on what individuals would do in any situations other than those observed. It might be noted, also, that the graphs displayed in both Girt [14] and Rushton [30] are inconsistent with either addition or multiplication. They are, however, consonant with the differentially weighted averaging model. As pointed out in this paper, current ordinal methods cannot handle such processes; hence, it seems likely that either their data are faulty or MDS cannot adequately describe the process. Functional measurement can describe such processes, and, if it can be shown to predict real spatial behavior, then it can derive models that are independent of unique spatial configurations. Thus, it potentially offers a more general approach to understanding spatial choice processes than is currently available.

As Anderson [3, 4, 6] has argued, the first order of business is the establishment of empirical laws relating independent factors to behavior. Such is also the case in geographic research, and it is hoped that the discussion in the paper can provide some direction toward this end.

REFERENCES

1. Anderson, N. H., 1972. "Cross-Task Validation of Functional Measurement," *Perception and Psychophysics*, 12, 389-95.
2. Anderson, N. H., 1972. "Looking for Configurality in Clinical Judgment," *Psychological Bulletin*, 78, 93-102.
3. Anderson, N. H., 1974. "Algebraic Models in Perception." In E. C. Carterette and M. P. Friedman (eds.), *Handbook of Perception* (vol. 2). New York: Academic Press.
4. Anderson, N. H., 1974. "Cognitive Algebra." In L. Berkowitz (ed.), *Advances in Experimental Social Psychology* (vol. 7), pp. 1-101. New York: Academic Press.
5. Anderson, N. H., 1974. "Cross-Task Validation of Functional Measurement Using Judgments of Total Magnitude," *Journal of Experimental Psychology*, 102, 226-33.
6. Anderson, H. H., 1974. "Information Integration Theory: A Brief Survey." In D. H. Krantz, R. C. Atkinson, R. D. Luce, and P. Suppes (eds.), *Contemporary Developments in Mathematical Psychology* (vol. 2), pp. 236-305. San Francisco: W. H. Freeman.
7. Anderson, N. H., 1976. "How Functional Measurement Can Yield Validated Interval Scales of Mental Quantities," *Journal of Applied Psychology* (vol. 61), 667-92.
8. Anderson, N. H., 1976. "Integration Theory, Functional Measurement, and the Psychophysical Law." In H. G. Geissler and Yu. M. Zabrodin (eds.), *Advances in Psychophisics*. Berlin: VEB Deutscher Vertag.
9. Birnbaum, M. H., 1974. "The Nonadditivity of Personality Impressions," *Journal of Experimental Psychology*, 102, 543-61.
10. Burnett, P., 1973. "The Dimensions of Alternatives in Spatial Choice Processes," *Geographical Analysis*, 5, 181-204.
11. Cadwallader, M., 1975. "A Behavioral Model of Consumer Spatial Decision-Making," *Economic Geography*, 51, 339-49.
12. Curry L., 1967. "Central Places in the Random Spatial Economy," *Journal of Regional Science*, 7, 217-38.
13. Dawes, R. M., 1971. "A Case Study of Graduate Admissions: Application of Three Principles of Human Decision Making," *American Psychologist*, 26, 180-88.
14. Girt, J. L., 1976. "Some Extensions to Rushton's Spatial Preference Scaling Model," *Geographical Analysis*, 8, 137-56.
15. Hensher, D.A., 1975. "Perception and Commuter Model Choice—An Hypothesis," *Urban Studies*, 12, 101-4.
16. Knight, R. L., and M. D. Menchik, 1976. "Conjoint Preference Estimation for Residential Land Use Policy Evaluation." In R. G. Golledge and G. Rushton (eds.), *Essays on the Multidimensional Analysis of Perceptions and Preferences*. Columbus, Ohio: Ohio State University Press.

17. Krantz, D. H., R. D. Luce, P. Suppes, and A. Tversky, 1971. *Foundations of Measurement* (vol. 1). New York: Academic Press.
18. Lerman, S. R., and J. J. Louviere, 1978. "On the Use of Functional Measurement to Identify the Functional Form of Travel Demand Models," a paper presented to the special session on Spatial Destination Choice, Transportation Research Board Meetings, Washington, D.C.
19. Lieber, S. R., 1977. "Attitudes and Revealed Behavior: A Case Study," *Professional Geographer* 79(7), 53-58.
20. Lieber, S. R., 1979. "An Experimental Approach for the Migration Decision Process," *Tijdschrift Voor Economische En Sociale Geografie*, 70 (2), 75-85.
21. Louviere, J. J., 1974. "Predicting the Evaluation of Real Stimulus Objects from an Abstract Evaluation of Their Attributes: The Case of Trout Streams," *Journal of Applied Psychology*, 58, 572-77.
22. Louviere, J. J., 1978. "Psychological Measurement of Travel Attributes." In D. A. Hensher and M. G. Dalvi (eds.), *Determinants of Travel Choice*. London: Tearfield, Farnborough (Saxon House Studies).
23. Louviere, J. J., and G. Kocur, 1979. "An Analysis of User Cost and Service Trade-offs in a Model Transit Demonstration Program." First Draft Final Report, TSC/UMTA Final Report Series, Cambridge Systematics, Inc., Cambridge, Mass.
24. Louviere, J. J., and R. J. Meyer, 1976. "A Model for Residential Impression Formation," *Geographical Analysis*, 8, 479-86.
25. Louviere, J. J., and R. J. Meyer, 1979. *Behavioral Analysis of Destination Choice: Theory and Empirical Evidence*. Iowa City, Iowa: University of Iowa, Institute of Urban and Regional Research, Technical Report #112.
26. Meyer, R. J., I. P. Levin, and J. J. Louviere, 1978. "Functional Determinants of Mode Choice," *Transportation Research Record* (Transportation Research Board, Washington, D.C.), fall, 1979.
27. Norman, K. L., 1973. *A Method of Maximum Likelihood Estimation for Information Integration Models*.San Diego: Center for Human Information Processing, Technical Report CHIP 37.
28. Norman, K. L., and J. J. Louviere, 1974. "Integration of Attributes in Public Bus Transportation: Two Modeling Approaches," *Journal of Applied Psychology*, 59, 753-58.
29. Piccolo, J. M., and J. J. Louviere, 1976. "Information Integration Theory Applied to Real-World Choice Behavior: Validational Experiments Involving Shopping and Residential Choice," a paper presented to the special session on Human Judgment, Meeting of the Association of American Geographers, Manhattan, Kans. October 1976. (Published in a special issue of the Great Plains/Rocky Mountains AAG: *Human Judgment and Spatial Behavior*, 1977, edited by R. Reider and J. Louviere.)
30. Rushton, G., 1969. "Analysis of Spatial Behavior by Revealed Space Preference," *Annals of the Association of American Geographers*, 59, 391-400.
31. Schmitt, R. P. "An Analysis of the Subdivision Site Selection Behavior of Residential Developers: Cedar Rapids and Des Moines, Iowa." Ph.D. dissertation, Department of Geography, University of Iowa, Iowa City, Iowa.
32. Shanteau, J., 1975. "Averaging Versus Multiplying Combination Rules of Inference Judgment," *Acta Psychologica*, 39, 83-89.
33. Slovic, P., and S. Lichtenstein, 1971. "Comparison of Bayesian and Regression Approaches to the Study of Information Processing in Judgment," *Organizational Behavior and Human Performance*, 6, 649-744.
34. Stutz, F., and R. A. Butts, 1976. "Environmental Trade-offs for Travel Behavior," *Professional Geographer*, 38, 167-71.

Part 3
Special Problems

Chapter 3.1 Some Remarks on Multidimensional Scaling in Geography

John S. Pipkin

The papers in this section on special problems fall neatly into two groups, a symmetry that I will disturb with some disjointed obser-vations on the explanatory role of MDS configurations and on a class of alternative representational models appropriate mainly in geographic applications.

Considering the diversity of perspectives on MDS techniques that has emerged in the social sciences and in marketing, the papers by Deutscher and Green are remarkably similar in outlook. They address the problems of data reliability associated with large stimulus sets. Deutscher's concerns encompass the merits of alter-native data collection procedures for stimuli too numerous for ex-haustive paired comparisons. Cross-sectional reliability can be assessed by smuggling repeated judgments into a single task. Longi-tudinal reliability is addressed as a problem of separating cross-sec-tional errors from changes induced by shifts in attitudes. Specific measures of reliability are proposed, and some substantive corre-lates of stability are evaluated. Green elaborates on Cliff's three criteria of reliability: internal consistency and replicability; struc-tural appropriateness (fit) of the scaling model; and the existence

The data on U.S. regions used in this study were gathered by Floyd M. Henderson, State University of New York at Albany. I am most grateful for his permission to use them and for informative discussions concerning them.

of the kind of external referents provided ideally by an embracive body of theory, and in practice by an ability to predict unobserved cases. Statistical criteria are elaborated by which these potential sources of unreliability may be unraveled.

These two papers testify to an emerging concern for data reliability that can only be addressed with interlocking and replicable data sets. They reveal the existence of such data (real and simulated) in impressive quantities. And they provide statistical techniques for evaluating reliability. Their substantive findings suggest that unreliability can be disturbingly great in data without at least partial replication.

In geography, as Deutscher observes, one dimension of reliability is far more tractable than in abstract attitudinal studies in which no external correlates of the imputed schemata are available. In studies that assess, say, the dimensions of retail desirability, the geographer's predicament does not differ materially from that of the market analyst. But, where the cognitive configuration of space itself is sought, as it is in the geographers' papers in this section, the objective configuration of cues provides the natural framework against which to measure stress, to assess learning convergence, and to study systematic distortions.

The papers by Golledge, Rayner, and Rivizzigno and Marchand adopt fit to objective space as their principal device for interpreting the recovered configurations. This stance permits structural inferences (e.g., the hypothesis on local distortions around primary anchor nodes) that would be hard to formulate and still harder to test in abstract attitudinal spaces without objective referents. Four distinct methods of performing such comparisons are demonstrated; two—hierarchic clustering and spectral analysis—are particularly powerful in disentangling local from global effects. The resulting analysis confirms the reliability of the recovered configurations according to the third and strongest of Cliff's criteria.

Marchand's comments address dimensionality. In a general case in which pairwise proximity estimates are elicited, dimensionality is unconstrained. An appropriate dimensionality is selected by examination of the profile of stress values for configurations recovered for various numbers of dimensions. A close fit with two dimensions and a Pythagorean metric is compelling evidence that the configuration converges on objective space [15]. Marchand addresses the case in which proximity measures are taken from sketch maps and rescaled. (This procedure seems to answer some of

MacKay's reservations on geographic MDS [27].) Marchand's second argument on fractal dimensions accords well with the finding that in some cases short distances are overestimated relative to longer ones [4]. Like tracing the length of smaller segments of seacoast on maps of increasing scale, "details keep popping up in the observer's mind as the path decreases in size."

Marchand's discussion of dimensionality highlights an unresolved problem in behavioral geography: how can MDS-recovered configurations be reconciled with "psychophysical" findings on the perception of distance between pairs of points? When all distances in a point set are transformed consistently according to the power or log functions that adequately describe pairwise distance perception [e.g., 4], the result has a determinate dimensionality in Euclidean space with the Pythagorean metric. Of course, the dimensionality would be higher than two. If we posit that learning results in a more nearly linear psychophysical function, the result would be a steady decrease in the dimensionality of the best-fitting MDS configuration. Learning or no, psychophysical accounts of distance perception have unexplored implications for MDS dimensionality.

In the remainder of this paper, two different classes of questions are raised that are not well integrated but that do seem pertinent to the present state of work in behavioral geography, where MDS techniques are virtually unchallenged as a means of eliciting global configurations of cognitive space. The first section questions or at least attempts to qualify the identification of MDS-recovered configurations with the internal schemata implied by the cognitive account of behavior. Tobler [43] indicates a line of least resistance in studying such schemata: it seems possible to make more definite statements on their geometry or relational structure than on their substantive content. The separability of structure from content, like the separability of knowledge and attitude, is itself an assumption that may, in some cases, be debatable because knowledge is motivationally coded [22], because intentions are governed by the selection of small subsets of salient attributes, and because points in geographic space possess multiple attributes. At any rate, MDS techniques have, in practice, exemplified this geometric orientation. The final section of the paper raises the question of eliciting and representing cognitive set structures; for example, the cognitive structure of neighborhoods, social areas, or regions on a national scale. Such representations would complement MDS configurations that characterize cues as points.

MDS AND THE EXPLANATION OF
GEOGRAPHIC BEHAVIOR

Cognitive Explanation

Most programmatic statements on the role of MDS-recovered configurations in behavioral geography imply a potential use as independent, causal, or explanatory constructs in "cognitive" models of overt spatial behaviors such as repetitive travel and residential mobility. But actual applications of MDS in explanatory modes are rare by comparison with studies that use the MDS configuration as a descriptive product. For example, several of the studies by Golledge and his co-workers employ the representations to illuminate aspects of cognitive content and structure, to compare individuals, and to portray learning, but not to explain specific observed behaviors [e.g., 15, 17].

It is inconceivable that study of elicited cognitive configurations should not yield insights on overt actions in, say, an urban setting. But it remains true that geographers have not devoted much attention to the formal, mathematical, or logical structure of such explanation or to the role of MDS configurations as explanans. The "cognitive" models that possess the strongest or at least the most transparent logical structure seem to be models of spatial choice in shopping travel or migration, which incorporate cognitive or subjectively scaled versions of traditional predictors, especially distance [e.g., 35]. The explanatory role of any construct evidently depends on the admissible structures of explanation. My objective in this section is to outline a rather restrictive account of a cognitive explanation and to appraise MDS-recovered configurations of geographic space from this point of view.

In psychology itself, the burgeoning of cognitive theory in recent decades has significantly complicated the explanation of behavior, at any rate by comparison with the rather transparent functional relations between stimulus and response sought in other psychological paradigms [29]. The term *cognitive* has apparently been used in a variety of ways [28]. Three key ideas are mediation, processing, and the causal remoteness (or even irrelevance) of immediate stimuli to overt behavior. Mediation posits an interrupted relationship between stimulus and response, as "some input process is categorized or related . . . to some more persistent internal process" [28, p. 187]. Internal schemata, including images or other cognitive representations, are commonly implicated. Processing emphasizes that information is used in a dynamic and reentrant fashion and serves to focus attention on propositional or other structural formats of schemata. The causal remoteness of action from context follows, in one temporal direction, from the persistence of schemata over time. In the other direction, it is implied by the cognitive premise that action is intentional in relation to a hierarchy of more or less remote goals.

The cognitive account of how the stream of ongoing behavior is governed stresses short- and long-term integrations as goals, schemata, and sense data are marshaled by the focus of conscious attention [10]. The subjective effect of these integrations is our sense of temporal continuity. Other implications of this account, confirmed by a moment's introspection, are the extremely limited purview and capacity of immediate attention and the almost complete inaccessibility of the causal processes and schemata, even to our own subjective scrutiny.

These apparent commonplaces on subjective experience yield a dense array of problems in experimental psychology, epistemology, "the psychophilosophy of knowledge" [34], and the philosophy of action. They also seem to yield a criterion for testing the relevance of explanatory constructs to ongoing spatial behavior in geographic contexts. To show causal or explanatory efficacy in a construct such as an MDS-recovered configuration, it seems necessary to show how the construct is related to the mediation of the stream of conscious or habitual action. This is an extremely tall order, considering that the actions geographers traditionally address are complex, are not subject to experimental control, and are inseparably fused with the totality of an individual's behavior. But to exhibit some of the mechanisms of this mediation would be a more convincing step toward cognitive explanation in geography than, say, a demonstration that subjective distance measures enhance the predictive power of an aggregate store choice model.

If indeed a thoroughgoing cognitive account of geographic behavior must exhibit connections between constructs and specific acts of overt behavior, it would seem easy to develop criticisms of many MDS studies in geography. Almost always, responses are elicited in laboratory settings as respondents perform (perhaps unaccustomed) discriminations and judgments over punctiform stimuli, which may be functionally mixed (nodes, landmarks, stores, theaters, etc.). Moreover, proximity judgments are usually elicited free from intentions and preferences, although intentionality (motivational coding) is central to the cognitive account of behavior [22].

Moroz, after reviewing various approaches to cognition, argues that the "representational function of cognitive processes is their most distinctive characteristic" [28, p. 205]. The most natural approach in appraising MDS configurations is to address the possible structural and functional similarities between the elicited representations and the imputed internal ones. Harman and Betak [19] provide an incisive attempt to do this from a geographic point of view. The only problem raised by these authors that seems to have receded in importance over the past few years is that of disaggregation. The adoption of programs such as INDSCAL and the application of scaling algorithms to recover configurations of individual respondents' spaces qualify Harman and Betak's statement that ". . . almost all empirical research reports results generalized for a group" [19, p. 13]. The comments below avoid the aggregation issue and focus on schemata on a strictly individual level.

Internal Schemata and Spatial Problem Solving

Geographers' statements on the form of internal cognitive representations disavow explicitly any assumptions on the physiological (cortical) form of the representations and concede that the representations need not be maplike in any simple sense [e.g., 11]. Gould [18] seems to go farther and implies a quasi-behavioristic position, which is simply to identify a "mental map" with an ability to make certain kinds of discriminations and judgments more or less easily.

Since facility in making judgments of various types may be related to the format of internal representation, it seems feasible to devise experiments capable of favoring one or another mode of internal representation. For example, spatial, linguistic, and pictorial modes have been shown to interfere in perceptual settings [12, 31, 39]. More pertinent to cognitive representations of space may be the fact that access to memory seems to be limited by the categorical structure of attributes that governed the original acquisition of the concepts [41]. There is strong evidence that mode of processing affects mode of encoding, and there is some evidence that deeper conceptual levels are more likely to have a semantic form [e.g., 26]. Judgments of a spatial type, epitomized by binary proximity judgments, can probably be made for many kinds of cues, considering the primordial nature of space as an organizing context and the prevalence of spatial metaphors even in language [45].

Nonetheless, preprocessing into a quasi-pictorial form seems necessary for such judgments to be made. Thus, judgment time might provide clues on the facility of such processing (e.g., from underlying semantic memories).

In addition, deductive considerations (e.g., capacity) permit plausible inferences on the organization of cognitive schemata of geographic space. It is inconceivable that astronomical numbers of pair-proximity estimates are accessible directly, but it is at least possible that places may be categorized according to regions ("proximity sets") anchored to key nodes. This hypothesis yields the testable implication that proximity judgments of pairs of places within sets may be superior to those between places in different sets. This seems to be supported by evidence that percepts are organized hierarchically, with global features having precedence over local ones. On a large geographical scale, Stevens and Coupe [42] show that locational judgments are made more readily within than between "superordinate" organizing units, such as the states of the United States.

Two general candidates for the internal organization of cognitive representations are the modes of propositions and imagery [30]. According to Rozeboom [34, p. 34], propositional knowledge is the "heartland of knowledge."The classical account of propositional knowledge can be characterized as follows: knowledge is mainly propositional, the predicates of propositions are categories (concepts), the truth values of propositions are beliefs, and the labels of concepts are mainly semantic. This portrait can be criticized from many points of view. For example, on a philosophical level, it seems that knowledge need not be propositional [33, p. 97]. And research on concept formation has complicated the simple set-theoretic account of concepts as categories [40].

Imagery seems a viable candidate for the organization of knowledge in cognitive as well as strictly perceptual contexts [20, 23, 24, 32]. Conceding the difficulties of operational definition, Kosslyn and Pomerantz [24] characterize images as nonperceptual, cognitively structured, spatially organized entities experienced by consciousness. Kosslyn and Pomerantz indicate experimental problems and possibilities in deciding between propositional and image forms of representations in various tasks.

The speed, facility, and interference of various kinds of judgments can in principle provide clues on the structural format of cognitive schemata, at least in the sense of second-order isomorphism. Superficially, this suggests that geographers might hope to impute, unequivocally, certain formats to certain schemata of space at a sufficiently deep level to provide unified explanations of laboratory judgments and overt behaviors. It would follow that MDS techniques might capture the structural relations in the unified schemata.

One class of difficulties in this program has not been discussed explicitly in the geographic literature. I will attempt next to outline its implications. The difficulties arise from the internal mutability of schemata, which are processed at various levels [8]. For example, transformations can link semantic and pictorial schemata [6]. The operation of such processes in geographic problem solving is conclusively shown when we consider, on one hand, the enormous complexity of cognitive schemata of space [22], and, on the other, the extremely limited capacity of attention and short-term memory. It seems certain that unknown processes tailor surface or experienced images as problems are consciously solved. A particularly economical way to think about space is to generate surface images that have many of the properties of visual percepts. In fact, spatial imagery has been shown to be evoked even in the solution of entirely nonspatial problems [21]. In such cases, conscious problem solving may proceed by reprocessing the surface image "by applying the same processes used in categorizing percepts and parts of percepts" [24, p. 70]. It seems certain that many geographic problems (both in urban travel and in laboratory settings) are solved in this manner.

It is possible to phrase problems that virtually coerce a given sequence of processes. Yi-Fu Tuan, for example, poses the question of recalling the stores in a well-known street [44, p. 62]. It seems that this unavoidably entails (1) generation of a visual surface image from deeper schemata and (2) reprocessing the surface image from a pictorial to a semantic mode. However, the bewildering variety of geographic problems solved in the course of everyday behavior suggests that it would be rash to dogmatize on the surface processes involved. Quite different surface processes are undoubtedly involved in, say, going home from work, giving street directions to a stranger, apartment hunting, and so on. Introspection suggests that one important distinction is between local problems solvable by eliciting one or a few strictly visual images of a locality (such as Tuan's problem above) and problems that involve comparisons over larger distances. Although it seems possible to elicit a long series of overlapping visual images of well-known parts of a city, thus allowing the linking of two distant locations, this seems a most unnatural way to judge their proximity. Moreover, substantial individual differences undoubtedly exist in, say, people's use of visual or verbal methods of problem solving. At minimum, a strictly cognitive account of spatial behaviors seems to require geographers to recognize the structural diversity of surface schemata and the variety of ways in which they are (consciously) processed in the course of problem solving.

MDS and Internal Schemata of Geographic Space

The discussion above suggests a model of geographic space cognition as a number of schemata linked by transformational processes. Deep structures (inaccessible to the conscious mind) may be transformed internally, for example, by the differential decay of memories of semantic or pictorial information. ("I can't remember the name but the place looks familiar.") Different transformations are abstracted from deeper structures to supply the conscious mind with surface schemata (including images). In turn, these schemata are consciously processed as specific questions are answered, decisions made, or problems solved.

The most distinctive feature of MDS techniques as applied in geography is their structural rather than process orientation. Evidently there is need for more direct investigation of the three levels of processing implied above (deep, deep to surface, conscious) as they pertain to knowledge and action in geographic space. Within the framework of structural applications, the discussion above implies the possibility of differing formats and relational structures of schemata at different levels. One obvious research strategy would be to attempt to make structural distinctions between them by using MDS techniques. Furthermore, surface schemata seem to be more accessible to this program than deep schemata. For example, we know that deep schemata are extremely complex, undoubtedly possessing imagelike, motor, semantic, and other features in associational networks of immense complexity; while surface schemata are, by definition, simple, more prone to be imagelike, and more accessible to introspection and experiment.

Unfortunately, it seems that MDS-recovered configurations cannot be *directly* interpreted as representing either surface images or deep schemata. Consider the paradigmatic case of a recovered configuration of points in urban space, elicited from an individual respondent and scaled, say, by KYST. The global nature of the recovered pattern belies the local and piecemeal way in which the respondent worked. The task extended over many minutes and involved the evocation and conscious reprocessing of many surface images. It would be absurd to identify the recovered pattern with any one surface image, even in the weak sense of second-order isomorphism. On the other hand, the configuration cannot be directly identified with a deep schema since the elicitation of surface images was colored by the form in which the subject's test was cast. For example, the decision between paired comparisons and rankings in urban distance estimates is usually considered an innocuous one governed by considerations such as convenience, respondents' fatigue, and the input required for particular scaling programs. Yet it is clear that subtly different surface processes are involved.

The mind's ability to tailor surface images to specific and very diverse tasks is a distinct liability in investigations of deeper schemata. An axiom of geographic applications is that it is meaningful to speak of geometric relational properties of underlying schemata, even though the format and the processing of surface images vary from spatial problem to spatial problem. This is logically necessary if lab-elicited schemata are to have any bearing at all on overt geographic behavior. If one accepts a priori the idea of a common set of deep-generating schemata for spatial relationships, the question becomes how invariant are their relational properties under the functional composition of two sets of transformations: deep-surface and surface reprocessing.

The structural orientation of MDS suggests that is is not well adapted to the study of the processes transforming the various levels of schemata at conscious or inaccessible levels. The study of strictly internal cognitive processes is not normally seen as necessary in process-oriented explanations in geography [but see 16]. The extent to which internal processes can be ignored undoubtedly depends on the extent of the invariances described above.

Up to this point, the explanatory problem of MDS configurations has been portrayed as one of relating hypothesized underlying schemata of space relationships to the surface processes of problem solving. The observation that many of the behaviors in repetitive travel are habitual [14] complicates the explanatory status of MDS configurations still further. All waking behavior is under cognitive control, but the cognitive operations of habitual travel seem more likely to involve template matchings (e.g., to monitor current location) than conscious judgments of spatial relationships. Such judgments (and the hierarchy of schemata and transformations they imply) are still implicated, but only during the comparatively short period during which habits were formed.

Four classes of research problems implied by these comments are the following. First, what are the surface (conscious) processes involved in solving spatial problems in natural or laboratory settings? Explicit recognition of the primacy of transient subjective processes in geographic behaviors promises to be useful in three ways. One is the pivotal role of consciousness that seems to be the central idea in cognitive explanation. One undeniable way in which internal schemata mediate behavior is in the surface images supplied to consciousness. A second: to assert that schemata have imagelike or maplike structure seems more defensible and certainly more testable in this context. And last, conscious experience is central to existential and phenomenological accounts of geographic behavior [5, p. 28]. A focus on this experience may provide one bridge between these philosophies and the profoundly positivistic stance of behavioral geography.

A second class of research questions is: how invariant are relational properties of surface images of real space under different formulations of the same abstract spatial problems (e.g., ranking versus paired comparisons)? The stress of an MDS-recovered configuration provides one measure of the global consistency of the numerous images elicited in the typical laboratory task. A single numerical measure of stress cannot capture systematic distortions arising, for example, when precision of distance estimates depends on distance. This would occur if visualizability of local areas of a city resulted in unusually precise short-distance estimates. Techniques discussed in the paper by Golledge, Rayner, and Rivizzigno (in this volume) would permit scale-dependent effects to be detected.

Third, is it meaningful in any sense to speak of invariant space relations in the deeper associational networks from which surface schemata are generated, and is it necessary or feasible for geographers to study them? Necessity is an empirical question of invariance. Feasibility is a question of developing experiments to distinguish, say, pictorial from semantic coding [1, 9].

Finally, does the structural orientation of MDS need to be complemented by studies of the three classes of internal processes mentioned above (deep, deep-surface, and surface)? This question, too, seems to be largely one of invariance in MDS configurations of geographic space.

TOWARD ALTERNATIVE REPRESENTATIONS
OF COGNITIVE SPACE

Whatever the difficulties of explicitly relating MDS configurations to overt spatial behavior, recent geographic work confirms their power in describing a series of spatial judgments in a concise and graphic way. The recovered metric configurations not only possess the strong heuristic appeal and descriptive economy of maps, they yield metric distance measures that are compatible with many traditional procedures, from gravity models of interaction within the cue set to numerical regionalizations.

One restrictive feature of most current applications, acknowledged for example by Golledge [15], is the requirement of punctiform cues. Logically, presented cues need not be identified as points in geographic space, but they are strictly pointlike in MDS output. Point cues in geographic space are commonly used not only in reconstructing schemata of space per se, but also in conjoint scaling of distance and place utility. The cognitive operations elicited from subjects—whether ordinal or metric, paired or ranked—essentially consist of point-point comparisons. Since the time of Lynch's work, it has been known that other topological entities than points figure in mental maps. By implication, many other cognitive operations than point-point judgments are involved in forming and processing schemata. Other types of judgments (such as point-set and set-set) can be studied in representations with setlike form.

The centrality of set structures and set operations in cognition and concept formation suggests that topological spaces may represent an appropriate language for a more general formulation of the MDS problem of reconstructing cognitive schemata and comparing them with objective geographic space. One conservative starting point is to concentrate on the set properties of the Euclidean plane, with Pythagorean or other metrics being ignored.

The possibility of inferring categorical structures to complement metric ones has been recognized in the multidimensional scaling literature [2, 3]. Boorman and Arabie, for example, appraise alternative metrics that can be defined on sets of partitions, where each partition represents categorical assignments made by one respondent. However, no literature addresses the problem of inferring cognitive set structures and comparing them with the objective structure of a system of sets defined in geographic space. Objective regional structures possess distinctive constraints (e.g., planarity) that need not be invoked in more abstract categorical problems such as those of semantics, kinship, and marketing [36]. The existence of many tools for the formal description of regional systems [7, 13], as well as evidence on cognitive regions [25], imply the possibility that the representation problem for cognitive sets in space may be both tractable and distinctively geographic.

Among the problems posed by this suggestion are the following. What are the salient cognitive properties of systems of regions (e.g., disjointness, containment, intersection, nesting, contiguity of various orders, and number of nearest neighbors)? How can such properties be elicited from subjects? How can representations of the implied cognitive schemata be developed (e.g., as cellular drawings; graphs such as the contiguity graph; matrices of the graphs, such as the adjacency, incidence, or cycle matrices)? How can the representations be compared with geographic space (e.g., what are suitable topological or graph-theoretic analogs of stress)?

Consider, for example, the following simple-minded problem of representing set-set judgments. Respondents are provided with named pairs of districts in a partition of some well-known area and asked to judge the contiguity (or, more weakly, the nearness) of each pair. Assuming forced symmetry, the resulting judgment matrix for any individual can be taken as in some sense approximating a contiguity matrix. This matrix might be realizable as a partition of the plane into simply connected and reasonably shaped districts. If some empirical contiguities were not elicited, the data might be representable as a set of disjoint but not exhaustive regions. On the other hand, the graph of the elicited contiguity matrix may not be planar, implying that no simple maplike representation is available. When point-point contacts are admitted as adjacencies (in addition to common borders), the representational problem becomes correspondingly less tightly determined. The problem of how an arbitrary symmetric zero-one matrix can be "most nearly" realized as a simple partition of the plane appears difficult, though the empirical configuration of districts would provide a natural starting configuration.

If the cue regions were not disjoint in real space, pair judgments could be solicited on disjointness, containment, and intersection. The representational problem in this case would involve seeking a Venn-like diagram in the plane, preserving as many as possible of the elicited set properties, with simply connected sets and with some constraint to preclude bizarre shapes.

The following empirical example illustrates a graph-theoretic approach to representing cognitive configurations and to comparing them with a corresponding objective structure. It concerns the cognition of large-scale regions of the United States, using data from respondents in three state universities in California, Louisiana, and Nebraska. Respondents in San Diego, Baton Rouge, and Lincoln were asked to assign the forty-eight conterminous states to nine regions specified only by name: East, Great Plains, Midwest, Northeast, Pacific Northwest, Rocky Mountain, South, Southwest, and West. Students were specifically instructed that states could be assigned to more than one region (i.e., the regions did not constitute a forced partition). The analysis reported here focused only on the role of empirical contiguity and higher-order adjacencies in constituting the regions. The cognitive contents of each region for each of the ninety-four usable responses were represented as subgraphs of the graph derived from the empirical contiguity structure of the states (a total of 846 representations). A similarity measure on [0,1] was defined between each pair of students (twice the following ratio: sum of assignments both students agreed upon/sum of all assignments made by both students). An exactly analogous similarity measure was defined between each pair of states.

The role of multiple-step adjacencies in constituting the regions was analyzed by regressing the aggregate state proximity measures for all distinct pairs of states on two independent variables. The first was shortest-path distance in the contiguity matrix of the forty-eight states ($r = -.59$). The second predictor was an aggregate measure of overall proximity: the normalized matrix of cumulated powers of the adjacency matrix, powered up to its diameter ($r = .20$). Shortest-path adjacency is a far better predictor of cognitive association between states than global connectivity. Residuals from these regressions reveal local (subregional) groups of states that are more or less strongly associated in respondents' minds than simple contiguity would predict. States of the South consistently possessed positive residuals, indicating strong cohesion, as did the states of the Northeast. Negative residuals indicated states that were less strongly associated with their neighbors, including, for example, Arizona, Arkansas, and Missouri.

For each respondent and for each region, two indexes of cohesion were computed: a zero-one indicator of connectedness and the number of isolated (unconnected) states (Tables 1 and 2). Substantial differences in cohesion of individual regions are evident. Cross-sample variations are less clear-cut. Of all possible combinations of samples and regions, only the following proportions were significantly different $(p < .05)$: Midwest (Louisiana/California), Northeast (Nebraska/Louisiana).

Several other analyses supported this evidence that variation within samples far outweighs any systematic effects associated with the three widely dispersed sample points. Various clustering routines applied to the individual proximity measures failed to reconstruct the three sample groups. The individual proximity matrix was subjected to principal components analysis and varimax and quartimax factor solutions. Patterns of loadings failed to distinguish

TABLE 1
Proportion of Connected Regions

Region	Nebraska Sample	Louisiana Sample	California Sample
East	.450	.588	.500
Great Plains	.700	.794	.590
Midwest	.700	.765	.500
Northeast	.250	.588	.395
Pacific Northwest	.950	.941	.850
Rocky Mountains	.800	.794	.892
South	.900	.853	.769
Southwest	.850	.818	.825
West	.850	.912	.725

TABLE 2
Mean Number of Isolates per Respondent

Region	Nebraska Sample	Louisiana Sample	California Sample
East	.600	.441	.625
Great Plains	.300	.235	.538
Midwest	.550	.412	.800
Northeast	1.250	.529	.921
Pacific Northwest	.150	.147	.225
Rocky Mountains	.300	.412	.216
South	.150	.353	.205
Southwest	.200	.242	.300
West	.300	.206	.475

the groups convincingly. The first principal component distinguished individuals primarily on the number of assignments made to regions (i.e., some respondents assigned many states to most regions; other students characterized many regions with few states).

Logically, responses assigning only a few states to a region might occur in two ways that should be distinguishable in the corresponding subgraphs of the global contiguity graph. The region might be perceived as a small core of contiguous states. Or the region might be characterized by a few typical, unusually salient or representative states, which would not necessarily be contiguous. This hypothesis would predict a bimodality in the frequency of isolates, controlling for number of states assigned. The hypothesis seemed plausible within the response set of some individuals, but no systematic effects were found across individuals or as the number of states assigned varied. Regressions of the number of connected regions against the number of states assigned produced weak associations that varied in magnitude and direction between the three sample groups and from region to region. Similarly, the number of isolated states assigned varied considerably across individuals but could not be convincingly associated in either a positive or a negative sense with the number of states assigned by each student. The frequency arrays of isolates per region across the ninety-four respondents peaked sharply at zero; Poisson distributions calculated from the mean frequency of isolates per respondent consistently underestimated low frequencies and overestimated high ones. A few individuals generated many isolates and disconnected regions but in a fashion not clearly related to the number of assignments made.

These partial and preliminary results can hardly be said to demonstrate the power of a graph-theoretic approach to recovering and representing the cognitive content of a regional system. The results show strong variations in the cohesion of different regions: the Pacific Northwest is the most cohesive, and the East and Northeast, the least cohesive. Significant differences between the sample groups were shown in some cases but not in most. The dominant feature of this data set seems to be unexplained individual variation in the connectivity of cognitive regions that has not yet been associated with identifiable cognitive styles in constituting them.

To represent the structure of cognitive regions as subgraphs of an objective adjacency structure does seem to be a fruitful way of attacking at least the limited issue of perceived contiguity, mainly because there exists a very extensive literature on describing the structural properties of graphs. The two indexes used above, an index of connectedness and the number of isolates, undoubtedly fail to describe adequately the full complexity of the cognitive regions. Current work with this data set addresses the problem of relating the structure of each individual's representation to the latent structure of objective contiguity. For example, when the shortest-path matrix is transformed from distancelike to correlationlike form and normalized, principal component analysis reveals a major structural dimension associated mainly with local degree (variance = 65% and subsidiary dimensions that appear to distinguish eastern states from western ones, and southeastern ones from all others. It remains to be seen whether these dimensions illuminate the structure of individual responses.

In an urban as well as a regional context, districts (neighborhoods, communities, social areas) are so central to social and cognitive organization that it seems highly desirable to develop appropriate methods for representing their cognitive structure. Not only are districts salient in their own right, but a network organization of cognitive spaces necessarily implies a dual structure of closed bounded regions with paths serving as edges. Golledge, Rayner, and Rivizzigno's model of node-path sets (see the next chapter in this volume) implies such a dual system of founded districts.

To elicit such cognitive structures requires identification of the pertinent judgments. To represent them as graphs (or to force them as nearly as possible into simply connected, reasonably shaped sets in the plane) requires nonmetric representations of cognitive schemata. Such representations would complement the metric products of conventional MDS.

REFERENCES

1. Bartram, D. J., 1974. "The Role of Visual and Semantic Codes in Object Naming," *Cognitive Psychology*, 6, 325-56.
2. Boorman, S. A., and P. Arabie, 1972. "Structural Measures and the Method of Sorting." In R. N. Shepard, A. K. Romney, and S. B. Nerlove (eds.), *Multidimensional Scaling* (vol. 1). New York: Seminar Press.
3. Boyd, J. P., 1972. "Information Distance for Discrete Structures." In R. N. Shepard, A. K. Romney, and S. B. Nerlove (eds.) *Multidimensional Scaling*, (vol. 1), pp. 213-23. New York: Seminar Press.
4. Briggs, R., 1973. "Urban Cognitive Distance." In R. M. Downs and D. Stea (eds.), *Image and Environment*, pp. 361-90. Chicago: Aldine.
5. Buttimer, A., 1976. "Grasping the Dynamism of Lifeworld," *Annals of the Association of American Geographers*, 66, 277-92.

6. Clark, H. H., and W. G. Chase, 1972. "On the Process of Comparing Sentences against Pictures," *Cognitive Psychology*, 3, 472-517.
7. Cliff, A. D., P. Haggett, J. K. Ord, K. Bassett, and R. Davies, 1975. *Elements of Spatial Structure: A Quantitative Approach*. Cambridge: Cambridge University Press.
8. Craik, F. I. M., and R. S. Lockhart, 1972. "Levels of Processing: A Framework for Memory Research," *Journal of Verbal Learning and Verbal Behavior*, 11, 671-84.
9. Desbarats, J. M., 1976. "Semantic Structure and Perceived Environment," *Geographical Analysis*, 8, 453-67.
10. Deutsch, J. A., and D. Deutsch, 1963. "Attention: Some Theoretical Considerations," *Psychological Review*, 70, 80-90.
11. Downs, R., and D. Stea, 1977. *Maps in Minds*. New York: Harper and Row.
12. Fox, L. A., R.E. Shor, and R. H. Steinman, 1971. "Semantic Gradients and Interference in Naming Color, Spatial Direction and Numerosity," *Journal of Experimental Psychology*, 91, 59-65.
13. Gale, S., and M. Atkinson, 1979. "On the Set Theoretic Foundations of the Regionalization Problem." In S. Gale and G. Olsson (eds.), *Philosophy in Geography*, pp. 65-107. Dordrecht: D. Reidel.
14. Golledge, R. G., 1970. "Some Equilibrium Models of Consumer Behavior," *Economic Geography*, 46, 417-24.
15. Golledge, R. G., 1978. "Representing, Interpreting, and Using Cognized Environments," *Papers of the Regional Science Association*, 41, 169-204.
16. Golledge, R. G., 1979. "Reality, Process, and the Dialectical Relation between Man and Environment." In S. Gale and G. Olsson (eds.), *Philosophy in Geography*, pp. 109-20. Dordrecht: D. Reidel.
17. Golledge, R. G., V. L. Rivizzigno, and A. Spector, 1976. "Learning about a City: Analysis by Multidimensional Scaling." In R. G. Golledge and G. Rushton (eds.), *Spatial Choice and Spatial Behavior*, pp. 95-116. Columbus, Ohio: Ohio State University Press.
18. Gould, P., 1976. "Cultivating the Garden: A Commentary and Critique on Some Multidimensional Speculations." In R. G. Golledge and G. Rushton (eds.), *Spatial Choice and Spatial Behavior*, pp. 83-91. Columbus, Ohio: Ohio State University Press.
19. Harman, E. J., and J. F. Betak, 1976. "Behavioral Geography, Multidimensional Scaling and the Mind." In R. G. Golledge and G. Rushton (eds.), *Spatial Choice and Spatial Behavior*, pp. 3-20. Columbus, Ohio: Ohio State University Press.
20. Holt, R. R., 1964. "Imagery: The Return of the Ostracized," *American Psychologist*, 19, 254-64.
21. Huttenlocher, J., 1968. "Constructing Spatial Images: A Strategy in Reasoning," *Psychological Review*, 75, 550-60.
22. Kaplan, S., 1973. "Cognitive Maps in Perception and Thought." In R. M. Downs and D. Stea (eds.), *Image and Environment*, pp. 63-79. Chicago: Aldine.
23. Kosslyn, S. M., 1976. "Can Imagery Be Distinguished from Other Forms of Internal Representation?" Evidence from Studies of Information Retrieval Time, *Memory and Cognition*, 4, 291-97.
24. Kosslyn, S. M., and J. R. Pomerantz, 1977. "Imagery, Propositions, and the Form of Internal Representations," *Cognitive Psychology*, 9, 52-76.
25. Lee, T. R., 1964. "Psychology and Living Space," *Transactions of the Bartlett Society*, 2, 11-36.
26. McDaniel, M. A., A. Friedman, and L. E. Bourne, 1978. "Remembering the Levels of Information in Words," *Memory and Cognition*, 6, 156-64.

27. MacKay, D. B., 1976. "The Effect of Spatial Stimuli on the Estimation of Cognitive Maps," *Geographical Analysis*, 8, 439-52.
28. Moroz, M., 1972. "The Concept of Cognition in Contemporary Psychology." In J. R. Royce and W. W. Rozeboom (eds.), *The Psychology of Knowing*, pp. 177-214. New York: Gordon and Breach.
29. Neisser, U., 1976. *Cognition and Reality*. San Francisco: W. H. Freeman.
30. Paivio, A., 1971. *Imagery and Verbal Processes*. New York: Holt, Rinehart and Winston.
31. Palef, S. R., 1978. "Judging Pictorial and Linguistic Aspects of Space," *Memory and Cognition*, 6, 70-75.
32. Pylyshyn, Z., 1973. "What the Mind's Eye Tells the Mind's Brain: A Critique of Mental Imagery," *Psychological Bulletin*, 80, 1-24.
33. Royce, J. R., 1972. "Comments on Moroz." In J. R. Royce and W. W. Rozeboom (eds.), *The Psychology of Knowing*, p. 209. New York: Gordon and Breach.
34. Rozeboom, W. W., 1972. "Problems in the Psycho-philosophy of Knowledge." In J. R. Royce and W. W. Rozeboom (eds.), *The Psychology of Knowing*, New York: Gordon and Breach.
35. Schuler, H. J., 1979. "A Disaggregate Store-Choice Model of Spatial Decision-Making," *Professional Geographer*, 31, 146-56.
36. Shepard, R. N., 1972. "A Taxonomy of Some Principal Types of Data and of Multidimensional Methods for Their Analysis." In R. N. Shepard, A. K. Romney, and S. B. Nerlove (eds.), *Multidimensional Scaling* (vol. 1), pp. 21-47. New York: Seminar Press.
37. Shepard, R. N., and S. Chipman, 1970. "Second-Order Isomorphism of Internal Representations: Shapes of States," *Cognitive Psychology*, 1, 1-17.
38. Shepard, R. N., A. K. Romney, and S. B. Nerlove (eds.) 1972. *Multidimensional Scaling* (2 vols.). New York: Seminar Press.
39. Shor, R. E., 1970. "The Processing of Conceptual Information on Spatial Directions from Pictorial and Linguistic Symbols," *Acta Psychologica*, 32, 346-65.
40. Simon, H. A., and K. Kotovsky, 1963. "Human Acquisition of Concepts for Sequential Patterns," *Psychological Review*, 70, 534-46.
41. Spyropoulos, T., and J. Ceraso, 1977. "Categorized and Uncategorized Attributes as Recall Cues: The Phenomenon of Limited Access," *Cognitive Psychology*, 9, 384-402.
42. Stevens, A., and P. Coupe, 1978. "Distortions in Judged Spatial Relations," *Cognitive Psychology*, 10, 422-37.
43. Tobler, W. R., 1976. "The Geometry of Mental Maps." In R. G. Golledge and G. Rushton (eds.), *Spatial Choice and Spatial Behavior*, pp. 69-81. Columbus, Ohio: Ohio State University Press.
44. Tuan, Y.-F., 1974. *Topophilia*. Englewood Cliffs, N.J.: Prentice-Hall.
45. Tuan, Y.-F., 1979. "Space and Place: Humanist Perspective." In S. Gale and G. Olsson (eds.), *Philosophy in Geography*, pp. 387-427. Dordrecht: D. Reidel.

Chapter 3.2 Comparing Objective and Cognitive Representations of Environmental Cues

Reginald G. Golledge, John N. Rayner,
and *Victoria L. Rivizzigno*

A survey of recent literature [23] indicates the existence of a considerable quantity of empirical evidence related to cognitions of different urban environments. Various investigations have pointed to the existence of point, areal, and linear components of cognitive representations of cities [1, 4, 7, 8, 20]. Since this research area is still in its infancy, however, considerable difficulties have arisen: first, concerning the extracting of information from individuals so as to develop cognitive representations; second, concerning attempts at incorporating the information thus extracted into maps or models of the representations. Some of these difficulties can be attributed to a lack of a viable conceptual framework and a theoretical basis for the studies [23]; additional problems are related to methods of representing and analyzing such representations. For example, very little attention has been paid to questions concerning the way that cognitive images of urban areas develop and change over time, the degree of similarity between recovered representations and real-world physical structures, the nature of biases and distortions as revealed in these representations, place-to-place and group-to-group variations in the type of representations

The research for this paper was supported by a National Science Foundation grant (GS-37969). We would like to acknowledge the assistance given by A. Spector and R. Zeller in the computer analysis of our subjects' responses.

recovered, and the relationship between the representations obtained from individuals and their different spatial behaviors. The purpose of this paper is to begin answering some of these questions, particularly by examining the relationship between configurations (cognitive representations) of a selection of environmental cues derived from multidimensional scaling procedures and cartographic representations (maps) of actual locations of the cues.

THE CONCEPTUAL FRAMEWORK

Tolman [37] has suggested that learning about an environment essentially consists of learning the location of places and the paths that connect them. Developmental psychologists such as Piaget [24] and Werner [38] argue that an adult's understanding and representation of space results from his or her interactions with elements in the physical environment rather than from perceptual copying of that environment. The interactional-constructivist point of view has proved to be a productive one, as evidenced by its increasing use by workers in environmental cognition. [For a detailed summary of the point of view and related work, see 23.] Other researchers [2, 10, 34] have drawn on both types of theories in presenting their own speculations about the development and structuring of environmental representations. The conceptualization that we adopt here fundamentally accepts the idea that learning about a large-scale environment results primarily from interacting with it, and more complex cognitive representations of these environments are built up over time. Like Tolman's, a position is adopted that individuals learn where certain places are in the environment and develop a variety of ways to get to and from and between the places. As more ways are explored, more places become known; as more places become known, there is a spread effect, or spillover, in the vicinity of those places, which results in the formation of knowledge about areas. Knowledge about places and their perceived significance can accrue from continued interaction (as with a place of work or residence) or because of social, historical, economic, ethnic, aesthetic, or other criteria of importance. Consequently, any individual's cognitive representation of, say, a city probably consists of points, lines, and areas that are connected to a greater or a lesser extent, depending on experience with the environment. These points, lines, and areas together form the "psychological mapping" that individuals compile of their environment.

At this stage, we must impress upon the reader that we are not necessarily suggesting that individuals carry node-path-area information in their minds in the form of Cartesian coordinates and Euclidean linkages or any similar type of recording mechanism. We do assume, however, that adult individuals have an understanding of spatial concepts such as proximity or closeness, dispersion, clustering, separateness, and orientation. In other words, it is assumed that spatial relations are understood, even if only implicitly, and that cognitive representations of external environments are based on each individual's interpretation of the elements of spatial relations. Of course, the precise interpretation that is given to each of these elements varies among individuals; such interpretations probably are also influenced by whether they involve relations among major cues (primary nodes), among secondary cues (or nodes), among primary and minor nodes, and so on. It is quite feasible that two primary nodes could be judged as being close in psychological space and that two minor nodes exactly the same physical distance apart could be judged to be quite far away in that same space. The essence of this conceptualization, however, is that, as places become better known to individuals, their spatial relations become better known. As spatial relations become better known, an individual more successfully integrates any new element of the environment into his or her representations (at the same time modifying his or her existing knowledge).

Given this emphasis on the spatial aspect of the cognitive representations, one can easily see how distortions and disturbances might occur in them. For example, during the early period of developing knowledge about an external environment, even the spatial relations among the more prominent cues may not be well understood; and, consequently, each of the latter may be misplaced or misconnected (i.e., misplaced or misconnected when compared to their positions or connections in objective reality). As primary nodes emerge from the general mass of environmental cues and as the spatial relations among the primary nodes become better known, segments of the environment become better known, as do the relations among them. For example, it is not too difficult to imagine hierarchical linkages between minor and higher-order nodes that considerably distort an individual's representational space because the primary node is offset or misplaced or disturbed in some way. Very likely, since minor nodes in the vicinity of a primary node would be tied very closely to it in a node-path link as well as in an areal sense, significant segments of the urban area could be disturbed because of the inaccurate rendering of the primary node.

As experience with an external environment (such as a city) increases, as familiarity with its elements increases, and as the hierarchical ordering of nodes and paths and areas becomes more firmly fixed in the mind of a person, a more coherent and cohesive representation of the external environment should emerge. Learning about such an environment then becomes a process of compiling information, which includes the spatial relations among environmental elements as well as their socioeconomic, cultural, or other meanings and significances. Over time, a person's cognitive representation should tend to become stable—at least insofar as its key elements (the primary nodes) are concerned. Once some type of stable representation has evolved, the nature of distortions and disturbances in it can be examined by comparing the stable representation with an objective representation, such as a two-dimensional map of the urban area.

Let us assume then that within any given macroenvironment (such as a city) there are a limited number of places that have a high probability of being defined as primary nodes by large subsets of the population of the area (and perhaps also by populations outside the area [22]). It should be possible to define a widely accepted primary node set for any such environment. Such a node set should provide the anchors for cognitive representations of the environment by a significant part of the population; as population members become more aware of the spatial relations among members of this node set, they should also be able to make judgments about them that reflect their knowledge of these relations. Individuals who are very well acquainted with a given primary node set should have little difficulty in reproducing their spatial relations (according to their own transformation); others less well acquainted with the node set may distort such relations in a variety of ways, but these distortions should become more consistent as the degree of knowledge increases. At this stage, therefore, it is feasible to hypothesize that the accuracy of the information about any given place or area is a function of the node-path set of the area (i.e., whether it contains primary or other nodes and, if so, how many and how they are connected) and the spillover effects from the node-path set that make up an individual's areal knowledge.

Thus, specific relationships between a given subarea of a city and the rest of the urban environment should depend on the subarea's node-path structure, and the accuracy of knowledge about the elements of an area should depend on the level of knowledge about its major nodes. For example, in an area where there is but a single primary node, other places may be referred to that node when being cognitively assimilated. If the dominant node is misplaced in some way, the likelihood of misplacement of subordinate places and the spatial relations among such places should be increased. As the degree of misplacement of the primary node changes, so too should the degree of misplacement of all other elements of the area change; and convergence may occur between the cognized relations and objective reality.

To summarize this conceptualization briefly, we are assuming that a large-scale external environment such as an urban area becomes known as a collection of parts. These parts consist of a hierarchical node set, a hierarchical path set connecting the nodes, and areas of various scales and with various degrees of generalization. The paths, nodes, and areas are linked together through the concept of spatial relations so that some type of comprehensive understanding of the large-scale external environment is obtained by the individual. It has not yet been suggested that this comprehension takes the form of any currently identifiable cartographic form or map projection; rather, it has been assumed that sufficient information will be known about where places are, how they are connected, and what their characteristics are to enable people to act "as if" they were capable of translating the information contained in their senses into some type of cartographic form. At this stage, there is very little information about the dimensionality of cognitive representations [30], very little information about whether representations can be stylized in the same way that maps of objective reality have been stylized, and very little information about the nature of biases that are built in to cognitive representations of different population subgroups. Work at Ohio State University since 1972 has concentrated on examining each of these questions in a limited urban environment (Columbus, Ohio); the balance of this paper is devoted to a discussion of the nature of some of these experiments and the results so far obtained from them.

RESEARCH DESIGN

The first step in operationalizing the preceding conceptualization is to determine whether a set of primary nodes exists within our sample area and, when they do exist, to itemize them. Given that such a set does exist, the next problem is that of discovering the nature of the cognitive configuration of these nodes. Further, we wish to see whether the accuracy of judgments made about spatial relations of the node set change over time as the level of knowledge about its members increases. The second stage (that of determining path sets) is not discussed in this paper. [See 8, 27.] The research design reported on here was developed in a pilot project at Ohio State University in 1972-1973 [10]. By drawing on a variety of published works and on extensive prior investigation in the city of Columbus, a number of types of environmental cues were selected. These cues were selected in the following ways: (1) on the basis of experimenter choice after detailed personal experience with the urban environment, (2) by eliciting responses from a number of students and faculty members at Ohio State University regarding the ten places "best known" to them in the city of Columbus, and (3) by contacting a number of persons outside the university atmosphere to obtain from them their ten "best-known" environmental cues.

The entire solicitation process resulted in approximately ninety cues being mentioned more than once by the sample population. Using a technique similar to that used by Milgram and associates [22], we simply recorded the frequencies with which each cue was mentioned by members of our sample population. In many cases, specific *places* (locations) in the urban environment were mentioned, with different *cue names* being used. A list of locations was compiled for the city, with one or more of the environmental cue names being used to define each location. Subsets of this list were then given to approximately 230 summer-school students at Ohio State University (this was a population somewhat more varied than a normal student population). Each subset has twenty locations; a variety of cue types were presented in each subset. Consecutive cue subsets had 75% overlap with previous subsets. (See Figure 1.) Subjects were first instructed to pick the place or places they knew best and give these a maximum scale score of 9 on a 9-point scale; then they chose the location(s) they knew least and gave them a score of 1. All other locations were to be given scores between 1 and 9 indicative of the relative amounts of knowledge or familiarity the subjects had of each of the cues and their locations. Next, subjects recorded the frequency with which they visited each place and their source of information about each place. The list of cue names was then reduced to those location-cue name combinations that were given high scores (either 8 or 9) on the 9-point scale. A simple ratio was then developed between the number of persons who gave scores of 8 or 9 for a location cue name and the number of people exposed to the location in the cue presentation schedule (Table 1). High ratios indicated that most of the people exposed to

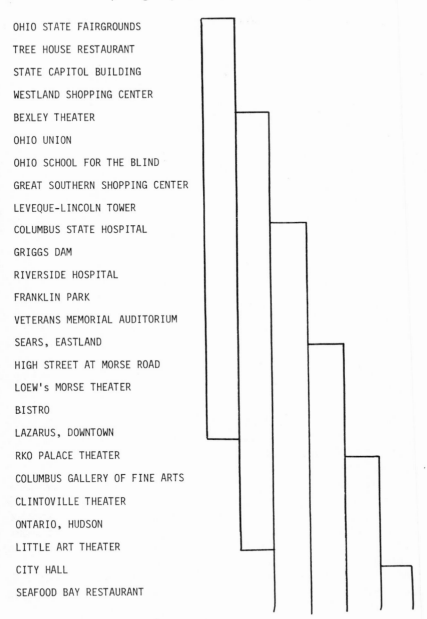

OHIO STATE FAIRGROUNDS

TREE HOUSE RESTAURANT

STATE CAPITOL BUILDING

WESTLAND SHOPPING CENTER

BEXLEY THEATER

OHIO UNION

OHIO SCHOOL FOR THE BLIND

GREAT SOUTHERN SHOPPING CENTER

LEVEQUE-LINCOLN TOWER

COLUMBUS STATE HOSPITAL

GRIGGS DAM

RIVERSIDE HOSPITAL

FRANKLIN PARK

VETERANS MEMORIAL AUDITORIUM

SEARS, EASTLAND

HIGH STREET AT MORSE ROAD

LOEW's MORSE THEATER

BISTRO

LAZARUS, DOWNTOWN

RKO PALACE THEATER

COLUMBUS GALLERY OF FINE ARTS

CLINTOVILLE THEATER

ONTARIO, HUDSON

LITTLE ART THEATER

CITY HALL

SEAFOOD BAY RESTAURANT

Figure 1. Cue selection procedures.

TABLE 1
Preliminary Test, Locations in Rank Order

Locations:	Familiarity Ratio:
1. Ohio Union	.866
2. Long's Bookstore	.810
3. Lane Ave. at Olentangy River Rd.	.795
4. Morse Rd. at 1-71	.785
5. OSU Football Stadium	.754
6. Gold Circle, Olentangy	.741
7. Ohio State Fairgrounds	.733
8. Northland Shopping Center	.732
9. Morse Rd. at Karl Rd.	.730
10. Broad St. at High St.	.674
.	
.	
.	
82. Thurber Towers	.050
83. Great Southern Shopping Center	.033
84. Bexley Theater	.033

the location cue name were very familiar with it. The twenty-four places with the highest ratios were then chosen as the subset of places most likely to be best known in Columbus; these constituted the primary location cue set for the major study (Table 2 and Figure 2).

Once locations were decided, each alternate and well-accepted environmental cue name for the location was collected. Since a dominant purpose of this study was to collect time-oriented data concerning knowledge of proximity of places (and other spatial relations), it was considered essential to minimize learning effects and order effects in the presentation of the cues to individuals. Learning effects were mitigated by presenting the same locations on consecutive trials but describing them by different (but approximately equally familiar) cue names. Not all locations were described in this manner; our prior experiments in this area had brought home to us the fact that individuals had varying degrees of familiarity with cue names and that this noticeably affected their judgmental processes. Consequently, some cue names were used consistently on *all* trials to provide anchors for comparative purposes (Table 2). Since many of the cue names were changed from trial to trial, it was felt that this also compensated for the order

TABLE 2
Location-Cue Sets

Cue Set 1	Cue Set 2	Cue Set 3	*
1. Ohio Union	Ohio Union	Ohio Union	(A)
2. Westland Shopping Center	Westland Shopping Center	Westland Shopping Center	(B)
3. OSU Football Stadium	OSU Football Stadium	OSU Football Stadium	(C)
4. Drake Union	Drake Union	Drake Union	(D)
5. Hudson St. at I-71	Hudson St. at I-71	Hudson St. at I-71	(E)
6. High St. at Morse Rd.	Graceland Shopping Center	Graceland Shopping Center	(F)
7. Ohio State Fairgrounds	Ohio State Fairgrounds	Ohio State Fairgrounds	(G)
8. Gold Circle, Olentangy	Gold Circle, Olentangy	Gold Circle, Olentangy	(H)
9. High St. at 15th Ave.	Long's Bookstore	Mershon Auditorium	(I)
10. Lazarus, downtown	Lazarus, downtown	Lazarus, downtown	(J)
11. Morse Rd. at Karl Rd.	Northland Shopping Center	Lazarus, Northland	(K)
12. A&P Supermarket at Hudson and High	Hudson St. at High St.	Hudson St. at High St.	(L)
13. Battelle Memorial Institute	Battelle Memorial Institute	Battelle Memorial Institute	(M)
14. Center for Science and Industry	Center for Science and Industry	Center for Science and Industry	(N)
15. Lane Ave. at Olentangy River Rd.	Lane Ave. at Olentangy River Rd.	Lane Ave. at Olentangy River Rd.	(O)
16. Leveque-Lincoln Tower	Broad St. at High St.	State Capitol Building	(P)
17. University Pharmacy	Lane Ave. at High St.	Arby's at Lane Ave. and High St.	(Q)
18. Park of Roses	Whetstone Park	Park of Roses	(R)
19. Morse at I-71	Morse at I-71	Morse at I-71	(S)
20. Veterans Memorial Auditorium	Veterans Memorial Auditorium	Veterans Memorial Auditorium	(T)
21. University City Shopping Center	Dodridge St. at Olentangy River Rd.	University City Shopping Center	(U)
22. Eastland Shopping Center	Sears, Eastland	Eastland Shopping Center	(V)
23. Greyhound Bus Station	Greyhound Bus Station	Greyhound Bus Station	(W)
24. Port Columbus Airport	Port Columbus Airport	Port Columbus Airport	(X)

*These letters correspond to the locations as they are printed out in the two-dimensional configurations.

effects of stimuli presentation, and no other steps were taken to eliminate this problem.

Subjects

It takes time to build a cognitive representation of the city for each individual—how much time we do not know exactly. Since it would be impractical to track an individual or a group throughout

TEST LOCATIONS, COLUMBUS, OHIO

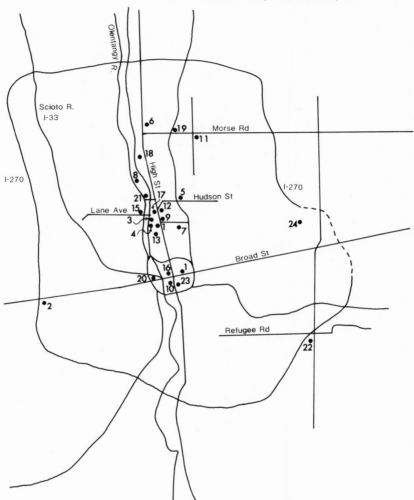

Figure 2. Location cues in Columbus, Ohio.

its life history to uncover the outcome of their information processing and recording mechanisms, we assumed that, in order to
cope with the exigencies of everyday living, some form of skeletal
node-path framework is built up by every urban individual and

that spatial modifications are made to this cognitive node-path set as information is continuously received about the external environment. In our prior conceptualization we also assumed that the preliminary node-path framework is built up by every urban individual; that spatial modifications are made to this cognitive node-path set as information is continuously received about the external environment; that the preliminary node-path set is probably tied to the activities of living, working, shopping, and recreating; and that we were capable of defining major places in each city that formed part of the skeletal cognitive image of the city for the majority of its individuals. Given these assumptions, it is reasonable to infer that long-time residents of an area and/or all populations acutely aware of the city as a whole would constitute a group who would know very well within the limits of their own cognitive transformation the relative position of well-known places (i.e., primary nodes). Such individuals can form a control group, which provides relatively stable representations of the selected cue locations.

It can further be assumed that others in the city will constantly modify their cognitive representations until some stable state becomes apparent. At this stage, we adopt the assumption that the stable-state representations of the previously described control group will approximate the asymptotes toward which other individuals in the city will strive. We might also note that the information that we recover from cognitive representations of the control group should allow us to determine whether our given environment is easy or difficult to cognize and to operate in [40]. Our population, therefore, consisted of a control group of longtime, well-informed residents, an experimental group of absolute newcomers to the city, and another experimental group of individuals who have some knowledge of the city but who may not have obtained the locational precision of the control group.

Sixty subjects were used in the data-collection phase; this group was broken down into three subgroups: (1) twenty-five city newcomers, (2) twenty-five intermediate-length residents, and (3) ten individuals in a control group. Each subgroup was further broken down into three subsets (a, b, c); this subdivision into smaller groups allows for the use of a Latin square design in the presentation of the cues.

The twenty-four locations were also divided into three groups, with a high degree of overlap between groups. Group I contained locations 1 to 9 and 22 to 24; group II contained locations 7 to 18;

and group III contained locations 16 to 24 and 1 to 3. Each cue was allocated to one of the three cue sets, with some cues being allocated to every cue set. Each subject group followed the same experimental design given in nine trials over a six-month period. For example, on trials 1, 4, and 7, subject subset *a* used cue set 1 and was given location group I; subject subset *b* used cue set 2 and was given location group II; subject subset *c* used cue set 3 and location group III. (Table 3 summarizes the complete cue-presentation schedule.) In this way, each group had confusion minimized by using only one cue set on all nine trials. But to prevent boredom and to minimize learning of places at each trial, they were given the mixes of locations in one of three subsets of the twenty-four locations.

TABLE 3
Cue-Trial Allocations

	Block 1	Block 2	Block 3
Trial 1, 4, 7	a 1 I	b 2 II	c 3 III
Trial 2, 5, 8	b 2 I	c 3 II	a 1 III
Trial 3, 6, 9	c 3 I	a 1 II	b 2 III

S_S = a, b, c; cue sets = 1, 2, 3.
Location groups = I, II, III.

ANALYTICAL METHODOLOGY

Earlier, we mentioned the problem of combining areal, linear, and point-located information into a single cognitive representation. To avoid this problem, we initially concentrated on specific places (points) in the data-collection phase. This allowed us to work with the "psychological distances" between places and to compare configurations derived from such distances with a map of the objective reality of the places.

Data-collection methods were designed to have a subject make a subjective judgment concerning the interpoint distances between pairs of points or to judge the degree of proximity between points. The inference made from such judgments was that at least ordinal-level data could be generated from the subjects' responses. Following Shepard [28, 29], Coombs [5], Kruskal [17, 18], and others interested in nonmetric scaling, we presumed that such ordinal-level data would contain a *latent spatial structure*. In other words, a set of interpoint distances could be generated so that the order of these distances corresponded to the order of the original judgments. Given a set of ordered distances, the task was then to find a space of minimum dimensionality so that, when the points were plotted in that space, the order of their interpoint distances was maintained. The two MDS routines used to obtain configurations of places from subjects were TORSCA-9 and KYST. Both of these are well documented [12, 19, 39] and widely used, so we will not discuss them here.

At the outset, we were undecided as to what algorithm to use and what starting configurations to choose. Therefore, we decided to compare the results obtained from each algorithm by using different starting configurations. This preanalysis procedure has been suggested by Spence [31, 32, 33] and others interested in comparing the efficacy of MDS algorithms. The results of this pretesting were presented by Golledge [8], and a final decision to use KYST as the major scaling tool was based on these results.

To ensure that each subject's configuration and the map of objective reality had the same potential orientation and the same scale, we subjected the matrix of objective distances between the twenty-four sample locations to the same algorithm (Figure 3). Various studies [10, 27, 41] have indicated that the goodness of fit between a configuration derived by using MDS and an original set of interpoint distances is excellent, with imperceptible distances or directional distortions. Configurations obtained from both the control group and the experimental groups thus can be compared directly to this MDS-derived map and "locational errors" for all two-dimensional cases can be calculated. "Locational errors" were calculated from Cartesian coordinates obtained after combining the locations of places on both the MDS representations of original interpoint distances and the subjects' configurations onto the same map [41].

EXPERIMENTAL RESULTS

Changes over Time

A number of fundamental hypotheses were to be tested at various stages of the analysis. One of these related to the process of environmental learning. We hypothesized that, if the fit between subjects' representations and objective reality increased over time, this would indicate that learning was proceeding. This hypothesis

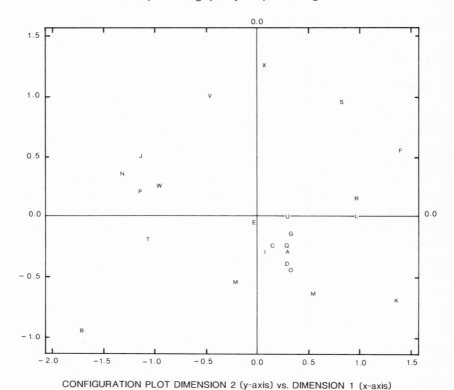

CONFIGURATION PLOT DIMENSION 2 (y-axis) vs. DIMENSION 1 (x-axis)

Figure 3. MDS configuration of location cues.

was tested in the following ways: (1) by examining changes in stress values, (2) by two-dimensional correlational analysis; (3) by hierarchical clustering analysis, and (4) by two-dimensional spectral analysis.

Stress Changes. Two major hypotheses were of concern with respect to the individual configuration stresses. First, it was hypothesized that, as a result of learning, the individuals who were initially new to Columbus (the newcomers' group) would be better able to locate the selected cues and more accurately define spatial relationships among them as time went on. It was expected that the stress values associated with the best configuration for each individual would decrease over time. (See Tables 4 and 5.) Using the Page statistic (a two-way analysis-of-variance test), the hypo-

TABLE 4
Newcomer Stress Statistics over Time

Subject	T_1	T_2	T_3
N1	.047	.010	.004
N2	.066	.082	.110
N3	.082	.005	.024
N4	.128	.148	.094
N5	.108	.095	.137
N6	.042	.041	.044
N7	.027	.012	.005
N8	.028	.060	.060
N9	.038	.005	.005
N10	.093	.034	—
N11	.079	.009	.002
N12	.085	—	—
N13	.092	.067	.032
N14	.092	.056	.062
N15	.047	.029	.010
N16	.121	.057	.052
N17	.100	.119	.045
N18	.049	.067	.106
N19	.078	.004	.071
N20	.134	.084	.051
N21	.005	.061	.004
N22	.100	—	—
N23	.035	.013	.008
N24	.063	.049	.005
N25	.037	.009	.005

SOURCE: Spector, 1975 [30].

TABLE 5
Page Statistics of Stress over Time,
Individuals Differentiated with Respect to Cue Set

Cue Set	Degrees of Freedom	L	Results
Q1	3,9	111.0	Cannot reject H_O with 95% confidence
Q2	3,7	91.0	Reject H_O with 95% confidence
Q3	3,6	79.5	Reject H_O with 95% confidence
All cue sets	3,22	281.5	Reject H_O with 99.9% confidence

SOURCE: Spector, 1975 [30].

thesis that no difference occurred over time was tested on data for the entire newcomers' group [30]. With 99% confidence, the null hypothesis was rejected (L(3,22 = 281.5), indicating that the proposed hypothesis was correct. When each cue set was differentiated by subgroup and the statistical test was performed on each, two of the three subgroups' statistics rejected the null hypothesis with 95% confidence. (See Table 5.) Thus, it would seem that with time the individuals tested could better handle the relations inherent in the positioning of the locations.

Stress values for the intermediate group showed no clear patterns of change over time (Table 6). The specific hypothesis tested was that stress values would decrease over time. Although mean stress values for the entire group did decrease somewhat, a test of

TABLE 6
Stress Values over Time, Intermediate Group
(KYST 3 Stress Values)

Subject	T_3	T_2	T_1
I1	.010	.025	.051
I2	.017	.010	.068
I3	.017	.050	.085
I4	.176	.125	.149
I5	.049	.010	.010
I6	.000	.010	.021
I7	.018	.020	.054
I8	.042	.066	.055
I9	.071	.138	.101
I10	.007	.036	.110
I11	.050	.075	.061
I12	.010	.007	.009
I13	.094	.069	.128
I14	.010	.010	.024
I15	.010	.009	.059
I16	.033	.014	.070
I17	.092	.180	.163
I18	.007	.012	.058
I19	.067	.109	.094
I20	.041	.086	.016
I21	.010	.008	.056
I22	.044	.042	.098
I23	.031	.017	.026
Means	.039	.049	.068

SOURCE: Brown, 1975 [3].

the significance of changes across all subjects using a one-way analysis of variance (Page statistic: $L = 2.4$) showed that no statistically significant change occurred [3]. Using a two-way nonparametric analysis of variance (the Friedman test), Thrall [35] also found that the probability that *no* change occurred was 90%. It appeared, therefore, that, though the newcomers' group showed considerable change in their ability to judge spatial relations among the chosen cues over time, the intermediate group made little, if any, progress.

It was assumed that the control group would have a relatively stable mapping of the spatial relations among cues over time, so they were tested only once. Their stress values are illustrated in Table 7. All the values are quite low, and visual inspection of the configurations indicated that they were readily interpretable, approximating the two-dimensional Euclidean map of the cue locations.

Correlational Analysis. Stress values by themselves are inadequate indicators of changes in the ability of individuals to comprehend the spatial relations among members of the location cue sets. But, just as the stress values changed significantly over time, so did the individual cognitive configurations, as indicated by the correlation between scale scores and locational coordinates of each cue. Two types of correlations were calculated to give a further idea of changes in the goodness of fit between objective and cognitive representations. Table 8 gives sample correlations for members of each population subgroup between the MDS-derived scale values for each cue set and its set of actual locational coordinates. Note that relatively little difference exists between sample subjects from the

TABLE 7
Control Group Stress Values

Subject Number	Stress$_1$
C1	.096
C2	.022
C3	.043
C4	.040
C5	.083
C6	.083
C7	.114
C8	.094
C9	.114
C10	.065

TABLE 8
Selected Correlations between Properties and Projections
(Profit)

Subjects	Time	N-S	E-W
Sample			
Newcomers:			
N3	T_1	.235	.449
	T_2	.418	.509
	T_3	.556	.776
N10	T_1	.493	.380
	T_2	.524	.719
	T_3	.419	.734
Sample			
Intermediates:			
I6	T_1	.386	.719
	T_2	.609	.594
	T_3	.554	.581
I3	T_1	.364	.474
	T_2	.305	.494
	T_3	.421	.691
Sample			
Control:			
C2		.951	.926
C4		.884	.926

SOURCE: Brown 1975 [3].

newcomers' and intermediate groups but that a significant difference exists between each of these groups and the control group. Spector [30] and Rivizzigno [27] further calculated two-dimensional correlation coefficients by using a configuration-matching procedure (CONGRU) [25]. The result of this testing for the newcomers' group and the control group are shown in Table 9, where it is quite obvious that the control group had achieved a far greater ability to match objective and subjective interpoint distances than had the newcomers. It appeared that when we accumulate all the subject populations and plot goodness-of-fit statistics against time, there is a tendency for judgments to improve quickly, level off, and then surge ahead again until some stable-state situation is achieved. It is at this later stage that a relatively stable and well coordinated cognitive representation of the external environment can be discerned.

Hierarchical Clustering. The configuration-matching procedure (CONGRU) gave useful and interesting results. However, we were

TABLE 9
Rho Squared Using CONGRU between the KYST
Configurations and the Actual Locations

Newcomers' Group	T_3	T_6	T_9
Cue Set I:			
N1	.507	.598	.901
N2	.719	.490	.386
N3	.659	.901	.884
N4	.210	.329	.503
N5	.283	.306	.378
N6	.738	.551	.820
N7	.224	.815	.959
N8	.808	.639	.868
N9	.557	.851	.858

Newcomers' Group	T_1	T_4	T_7
Cue Set II:			
N10	.266	.428	
N11	.594	.908	.904
N12	.346	.520	.576
N13	.182	.627	.774
N14	.797	.885	.861
N15	.242	.755	.780
N16	.256	.509	.607
N17	.584	.633	.607
N18	.397	—	—

Newcomers' Group	T_2	T_5	T_8
Cue Set III:			
N19	.497	.356	.682
N20	.672	.748	.514
N21	.653	.663	.870
N22	.475	—	—
N23	.821	.931	.813
N24	.534	.606	.750
N25	.699	.931	.969

Control Group	T_1
C1	.891
C2	.952
C3	.855
C4	.944
C5	.938
C6	.882
C7	.889
C8	.901
C9	.709
C10	.882

still not satisfied; therefore, we decided to use a clustering approach to help determine whether or not individual and group configurations closely approximated the actual location set. To achieve this, a hierarchical clustering algorithm, HICLUST [16], was applied to the data. As opposed to the vector-fitting procedures of CONGRU, the hierarchical clustering algorithm showed graphically the spatial organization contained within both the actual locations and the derived configurations. Clustering analysis tries to uncover the basic stimulus organization by forming groups of stimuli so that intragroup similarity is maximized while intergroup similarity is minimized.

The data for HICLUST consisted of the interpoint distances between points on both the MDS-derived configuration and the map of actual locations. Clustering was performed for each subject and the actual locations; Figure 4 shows the results of using the diameter method for clustering the actual locations, and Figure 5 shows the related dendrogram. The dendrogram for the diameter method shows various groups of locations that seem to be free of the chain

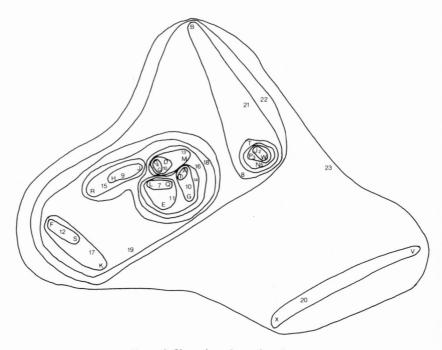

Figure 4. Clustering of sample points.

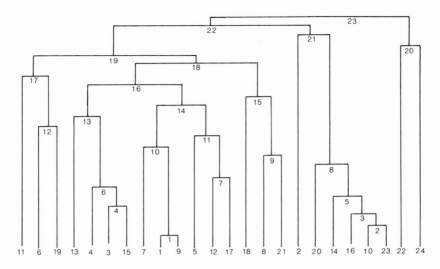

Figure 5. Dendrogram for cluster map.

effect. The basic clustering units appear to be (1) locations in the extreme (22, 24); (2) downtown locations, joined later by that location in the extreme west; (3) campus area locations; and (4) far-north locations.

The control group's results when the diameter methods were used showed some clusters that did not appear when the actual locations were analyzed, but there was in each case a high degree of correspondence between the two [27].

When the diameter method was used for the newcomers, each configuration was divided into four groups and contingency tables were constructed by utilizing criteria that would maximize the similarity between the grouped actual locations and the analogous grouped configuration locations. Table 10 indicates that between 59% and 76% of the points in the KYST configurations could be grouped analogously with the actual locations. More important, the percentage seemed to increase over time (the Page statistic, $L = 41$ with $df = 3, 3$ is significant with 95% confidence). For differences among cue sets, the Friedman test indicates that the null hypothesis of no difference among cue sets cannot be rejected with 95% confidence (chi-square $[df = 2] = 4.667$).

As in previous cases, the configurations for the intermediate group were more difficult to analyze. However, again it was shown

TABLE 10
Contingency Tables Based on Grouping of
Actual Locations and KYST Configurations Using HICLUS

$Q1\ T_3$
Configuration/Actual Locations

	1	2	3	4	Total	
1	19	4	0	4	23	
2	4	32	0	0	41	
3	3	9	6	5	17	
4	1	9	3	1	25	
Total	27	54	9	12	108	% Correct: 64.3

$Q1\ T_6$
Configuration/Actual Locations

	1	2	3	4	Total	
1	22	4	0	0	26	
2	3	36	0	7	46	
3	1	4	9	6	20	
4	1	10	0	5	16	
Total	27	54	9	18	108	% Correct: 66.4

$Q1\ T_9$
Configuration/Actual Locations

	1	2	3	4	Total	
1	24	4	0	0	26	
2	0	38	1	4	43	
3	2	2	7	6	17	
4	1	13	1	8	23	
Total	27	54	9	18	108	% Correct: 71.3

$Q2\ T_1$
Configuration/Actual Locations

	1	2	3	4	Total	
1	21	10	0	2	33	
2	4	28	2	6	40	
3	1	6	7	1	15	
4	1	10	0	9	20	
Total	27	54	9	18	108	% Correct: 59.3

$Q2\ T_4$
Configuration/Actual Locations

	1	2	·3	4	Total	
1	21	4	0	0	25	
2	2	35	2	9	48	
3	1	4	5	0	10	
4	0	5	7	7	13	
Total	24	48	8	16	96	% Correct: 70.8

TABLE 10—*Continued*

Q2 T₇
Configuration/Actual Locations

	1	2	3	4	Total
1	18	5	0	0	23
2	1	27	0	5	33
3	0	1	7	1	9
4	2	9	0	8	19
Total	21	42	7	14	84

% Correct: 71.4

Q3 T₂
Configuration/Actual Locations

	1	2	3	4	Total
1	15	3	0	1	19
2	3	33	0	4	36
3	0	1	6	1	8
4	3	5	1	8	18
Total	21	42	7	14	84

% Correct: 73.8

Q3 Tₛ
Configuration/Actual Locations

	1	2	3	4	Total
1	16	3	0	0	19
2	2	24	0	3	29
3	0	0	6	0	6
4	0	9	0	9	18
Total	18	36	6	12	72

% Correct: 76.4

Q3 T₈
Configuration/Actual Locations

	1	2	3	4	Total
1	16	3	0	0	19
2	1	23	0	3	27
3	1	0	6	1	8
4	0	10	0	8	18
Total	18	36	6	12	72

% Correct: 73.6

that a high degree of correspondence could be obtained between the actual clusterings and the clusters based on individual configurations.

Spectral Analysis. Analysis of the MDS-derived configurations via cartographic methods, configuration-matching procedures, and clustering were next supplemented by a more detailed discussion of the comparative structuring of actual and derived configurations using two-dimensional spectral analysis. In this study, the MDS-derived configurations are based on a set of selected best-known places. As such, the locations act as first-order points—the pegs on which the rest of the structure can be hung. The analogy with geodetic surveying is a close one, and, indeed, Tobler [36] has already recognized that relationship and exploited it in developing techniques for analysis. The analogy should not be carried too far, however, because emphasis is placed on different aspects of the triangular points. In geodetic surveying, no questions are asked about the reasons for measured distances and orientation. Only the magnitudes and the precision of the measurement are of interest. In contrast, in the study of cognitive structures the analysis goes beyond that step and comparisons with objective reality are made in an attempt to understand and explain the distances and orientations.

One approach to this problem, as demonstrated above, is to fit a rectangular grid from objective reality to the individual configuration (this procedure is discussed in the final section of this paper). The output maps indicate local and regional distortions. Some limitations of such maps are apparent. One "impaired" cue can change the whole structure of a grid map because it can affect both east-west and north-south grid lines. Also, the character of the map can be controlled by the shape of the peripheral areas that contain no cues. In part, this is due to the assumption that the region is composed of a continuous surface and that interpolations of various types are possible. An alternative method of analysis suggested here is one that investigates the spanning of cues without reference to the character of the intervening space. This is spectral analysis.

Spectral analysis as a method of analyzing the spacing of location in two-dimensional Euclidean space has been discussed by two of the authors elsewhere [26]. Hence, only a brief summary will be given here. Spectral decomposition is a necessary component of statistical analysis whenever there is an interdependence of the data as a result of arrangement, such as in space or in time. Thus, a normally distributed spatial series requires the mean and variance spectrum for its characterization. Similarly, for two normally distributed and related spatial series, the regression and correlation coefficients should be replaced by the phase, gain, and coherence spectra. A spectrum is another way of looking at autocorrelation and lag-correlation: one is the Fourier transform of the other. As it turns out, the spectrum is easier to understand, and its confidence bands are more readily calculated.

In order to explain the spatial spectrum, the statistic variance will be used. The variance spectrum is the scale decomposition of the variance. It shows how important (proportionally) each size of feature is in contributing to the total variance. For a two-dimensional array there is a further decomposition according to orientation. Because the variance in one orientation is identical to that in the reverse orientation, only half the output spectrum need be shown. The largest scales (lowest frequencies) appear in the center of the top line of the figure, with smaller and smaller scales (higher

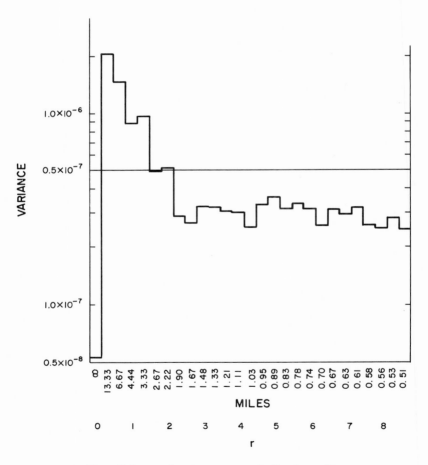

Figure 6. Averaged vector spectrum of objective reality.

frequencies) appearing radially from that point. Sometimes it is useful to assume that the array of original data was isotropic. Then, orientation can be ignored, and the two-dimensional spectrum can be collapsed to one by averaging in arcs of constant scale (frequency). The result has been called the averaged vector spectrum of variance. (See Figure 6.)

Because the particular technique used in this paper requires equally spaced observations, a fine grid has been established so that each environmental cue falls exactly on, or close to, integer coordinates. Then, the matrix of integer intersections are set to zero, except those coinciding with the data. These are set equal to one. In such an array, the total group of urban cues will be the largest scale feature; and twice the grid cell size, the smallest. Scales relating to cues that fall on adjacent grid points will not be resolved. The variance associated with them will be incorporated into the lower frequencies through a process known as "aliasing." Consequently, a grid was chosen so that there is always at least one zero point between each pair of cues. The grid spacing in both north-south and east-west directions is approximately 0.07 miles. This limit, together with the overall size of the area (14 x 14 miles), gives good resolution at scales less than 1 mile but poor separation at the larger scales, where most of the pattern seems to emerge. The objective locations in future studies should be selected to remove this difficulty. The scales (or wavelengths) in the center of the spectral bands are listed on the averaged vector spectra. The scales for the two-dimensional spectra are given in Table 11.

TABLE 11
Scales (Wavelengths) at the Center of the Spectral Bands in Miles

		0	1	2	3	r_1 4	5	6	7	8
	0		4.44	2.22	1.48	1.11	0.89	0.74	0.63	0.56
	1	4.44	3.14	1.99	1.41	1.08	0.87	0.73	0.63	0.55
	2	2.22	1.99	1.57	1.23	0.99	0.83	0.70	0.61	0.54
	3	1.48	1.41	1.23	1.05	0.89	0.76	0.66	0.58	0.52
r_2	4	1.11	1.08	0.99	0.89	0.79	0.69	0.62	0.55	0.50
	5	0.89	0.87	0.83	0.76	0.69	0.63	0.57	0.52	0.47
	6	0.74	0.73	0.70	0.66	0.62	0.57	0.52	0.48	0.44
	7	0.63	0.63	0.61	0.58	0.55	0.52	0.48	0.44	0.42
	8	0.56	0.55	0.54	0.52	0.50	0.47	0.44	0.42	0.39

Although the individual configurations are of primary interest, it is necessary initially to analyze the MDS configuration of the actual locations. This is because the location of the cues in objective space represents an organized pattern that has its own signature spectrum. An individual's configuration is some transformation of that pattern. Unfortunately, at this time we cannot remove the influence of that particular objective pattern on an individual's cognitive processes. Certain objective patterns may be more easily transformed than others.

Figure 7 gives the central 17 x 17 bands or blocks of the two-dimensional spectrum of the objective reality locations. Higher frequencies are not plotted in any of the outputs discussed below, even though they were calculated, because they show that the finer aspects of the location pattern were random. Figure 7 indicates that only at scales larger than 2.5 miles did any clear nonrandom pattern occur in the objective locations. A significant peak at $r_1 = 1$, $r_2 = 0$ is the result of the cues being concentrated east of a north-south line just west of the city center. The general grouping of high variance around the origin is due to the concentration of cues in the central region. This is borne out by the averaged vector spectrum (Figure 6), which shows a peak of variance at lengths greater than 2.5 miles. Higher frequencies fluctuate around a constant variance of 0.3×10^{-7} units in a manner to be expected from random data.

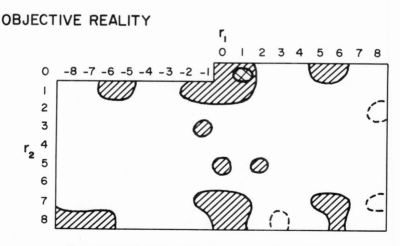

OBJECTIVE REALITY

Figure 7. Two-dimensional spectrum of objective reality.

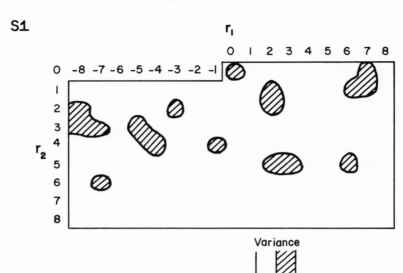

Figure 8. Two-dimensional spectrum for subject S1.

To illustrate the results that are obtained from the spectral analysis, two individual configurations have been selected. They were taken from the control group, whose members, theoretically, knew the city well. The first of these, S1, produced an almost random configuration as estimated by the two-dimensional spectrum (Figure 8). Here the poor resolution at the lower frequencies (larger scales) may obscure a pattern. The averaged vector spectrum, which does have higher resolution, indicates that on the general city scale there appears to be some nonrandom arrangement. However, it matches the objective pattern only at the largest of these scales (Figure 9). The two-dimensional coherence spectrum shows poor agreement everywhere, but the average vector spectrum gives coherences (analogous to the coefficient of determination) of 0.98 and 0.75 at 13.3 and 6.67 miles, respectively. Thereafter, the coherence remains at less than 0.15. It can be said, then, that S1 was able to locate the cues at the largest scales only and that he or she was unable to reproduce reality at scales less than at about 5 miles. The distinct north-south line was not recognized by S1.

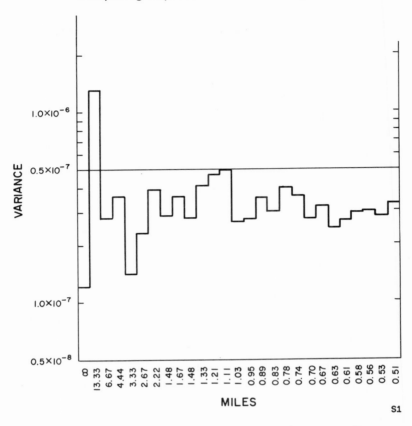

Figure 9. Averaged vector spectrum for subject S1 (from Figure 8).

S2 appears at the other extreme. The two-dimensional spectrum (Figure 10) reveals a nonrandom pattern at relatively small scales (down to 1 miles). Also, an orientation is present that indicates that the north-south separation mentioned above is rotated slightly to NNE-SSE. Again, the averaged vector spectrum (Figure 11), although less valuable for a pattern that contains a definite orientation, shows significant variance at wavelengths greater than 1 mile. The relationship with the objective pattern is good down to about 2.5 miles. The two-dimensional coherence around the origin is given in Table 12. It shows that S2 has a configuration that matches very well the objective location, where they are arranged in a

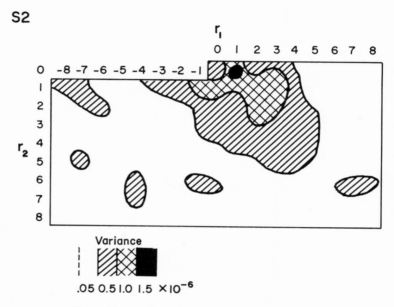

Figure 10. Two-dimensional spectrum for subject S2.

nonrandom fashion. The averaged vector spectrum produces coherences of 1.0, 0.99, 0.83, 0.80, 0.72, and 0.25 over the first six
bands. (See Figure 11 for wavelengths.) Thereafter, the relationship is poor (less than 0.20).

In summary, S2 has a highly structured configuration when compared to objective reality. The cues are located in a nonrandom
pattern down to the 1-mile scale. In addition, there is a clearly defined orientation in that pattern. This structure matches very well
objective reality, where the objective reality has a nonrandom pattern (greater than 2.5 miles).

In conclusion, it can be stated that spectral analysis of configuration data produces several types of valuable results beyond those
required for statistical description. It permits the recognition of
random and nonrandom elements in the locations on the basis of
scale. It distinguishes between individual configurations and allows
for the classification of individuals in terms of the extent to which
they structure locations. It shows how well individual configurations match objective reality at different scales.

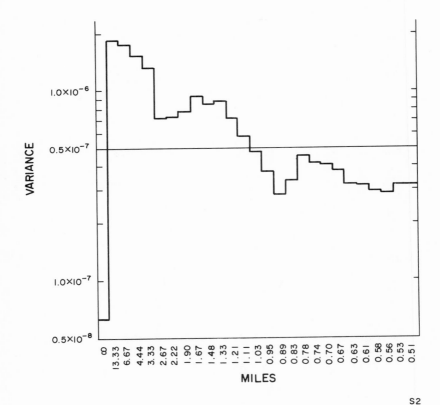

Figure 11. Averaged vector spectrum for subject S2 (from Figure 10).

SUMMARY

In this paper we have stressed a conceptual model, an experimental design, and a set of analytical results. The latter are still being compiled for the bulk of the sample members. This rather brief analysis has raised the following questions.
1. What *is* it that people learn about urban areas? And how quickly do they learn it?
2. Does the mode of transportation used influence cognition of spatial relations?

TABLE 12
Coherence (R^2) between the Objective Locations and the S2 Configuration

				r_1		
				0	1	2
r_2	0			0.99	0.90	0.15
	1	0.05	0.51	0.68	0.82	0.67
	2	0.17	0.03	0.22	0.64	0.49

3. How does access to various information sources influence cognized spatial relations?
4. What level of information about a place is required before its relative positions are known?
5. What *produces* the holes in cognitive representations and the various distortions of spatial relations in those representations?
6. Do well-known areas (such as neighborhoods) *seem* larger than they are? And what significance has this for the planning and the design of new towns and old cities?
7. What is the mathematical function used to translate physical distance into scaled responses?

As answers to these and many other questions are determined, our overall picture of what people learn about cities, why they learn those things, and what use they make of learned phenomena will become clearer. As this picture becomes clearer, the linkages between explanatory variables will become more evident, and the way will be cleared for the integration of experimental results into useful and meaningful theory concerning the cognitive structure of large-scale environments.

REFERENCES

1. Appleyard, D., 1969. "City Designers and the Pluralistic City." In L. Rodwin and associates (eds.), *Planning Urban Growth and Regional Development: The Experience of the Guyana Program of Venezuela*, pp. 422-52. Cambridge, Mass.: M.I.T. Press.
2. Appleyard, D., 1970. "Styles and Methods of structuring Cities: Part Two," a paper presented at the Symposium on Conceptual Issues in Environmental Cognition, Environmental Design Research Association Meeting, Los Angeles, 1970.
3. Brown, M., 1975. "Cognitions of Distance in a Metropolitan Area: The Intermediate Group." In R. G. Golledge (ed.), *On Determining Cognitive Configurations of a City*, pp. 333-48. Columbus, Ohio: Department of Geography, Ohio State University Research Foundation.

4. Carr, S., and D. Schissler, 1969. "The City as a Trip: Perceptual Selection and Memory in the View from the Road," *Environment and Behavior*, 1, 7-35.
5. Coombs, C., 1964. *A Theory of Data*. New York: John Wiley and Sons.
6. Cox, K. R., and R. G. Golledge (eds.), 1969. *Behavioral Problems in Geography*. Evanston, Ill.: Department of Geography, Northwestern University, Studies in Geography #17.
7. Craik, K. H., 1970. "Environmental Psychology." In K. H. Craik et al. (eds.), *New Directions in Psychology*, (vol. 4), pp. 1-121. New York: Holt, Rinehart and Winston.
8. Golledge, R. G., 1975. *On Determining Cognitive Configurations of a City*. Columbus, Ohio: Department of Geography, Ohio State University Research Foundation.
9. Golledge, R. G., R. Briggs, and D. Demko, 1969. "The Configuration of Distances in Intraurban Space," *Proceedings of the Association of American Geographers*, 1, 60-65.
10. Golledge, R. G., and V. L. Rivizzigno, 1973. "Learning about a City: Analysis by Multidimensional Scaling," a paper presented to 69th Annual Meeting of the Association of American Geographers, Atlanta, Georgia.
11. Golledge, R. G., V. L. Rivizzigno, and A. Spector, 1975. "Learning about a City: Analysis by Multidimensional Scaling." In R. G. Golledge (ed.), *On Determining Cognitive Configurations of a City*. Columbus, Ohio: Department of Geography, Ohio State University Research Foundation.
12. Golledge, R. G., and G. Rushton, 1973. *Multidimensional Scaling Review and Geographical Applications*. Washington, D.C.: Association of American Geographers, Commission on College Geography, Technical Paper #10.
13. Golledge, R. G., and G. Rushton (eds.), 1976. *Spatial Choice and Spatial Behavior*. Columbus, Ohio: Ohio State University Press.
14. Golledge, R. G., and G. Zannaras, 1973. "Cognitive Approaches to the Analysis of Human Spatial Behavior." In W. H. Ittelson (ed.), *Environment and Cognition*, pp. 59-94. New York: Seminar Press.
15. Isaac, P., and D. Poor, 1974. "On the Determination of Appropriate Dimensionality in Data with Error," *Psychometrika*, 39, 91-109.
16. Johnson, S., 1967. "Hierarchical Clustering Schemes," *Psychometrika*, 32, 241-54.
17. Kruskal, J., 1964. "Multidimensional Scaling: A Numerical Method," *Psychometrika*, 29, 115-219.
18. Kruskal, J., 1964. "Multidimensional Scaling by Optimizing Goodness-of-Fit to a Nonmetric Hypothesis," *Psychometrika*, 29, 1-27.
19. Kruskal, J., F. Young, and J. Seery, no date. "How to Use KYST, a Very Flexible Program to Do Multidimensional Scaling and Unfolding." Unpublished manuscript, Bell Laboratories, Murray Hill, N.J.
20. Lynch, K., 1960. *The Image of the City*. Cambridge, Mass.: M.I.T. Press.
21. Milgram, S., 1970. "The Experience of Living in Cities," *Science*, 167, 1461-68.
22. Milgram, S., J. Greenwald, S. Kessler, W. McKenna, and J. Waters, 1972. "A Psychological Map of New York City," *American Scientist*, 60, 194-200.
23. Moore, G., and R. G. Golledge, (eds.), 1976. *Environmental Knowing*. Stroudsburg, Pa.: Dowden, Hutchinson and Ross.
24. Piaget, J., 1954. *The Construction of Reality in the Child*. New York: Basic Books.
25. Olivier, D., 1970. "Metric for Comparison of Multidimensional Scaling." Unpublished manuscript.
26. Rayner, J. N., and R. G. Golledge, 1972. "Spectral Analysis of Settlement Patterns in Diverse Physical and Economic Environments," *Environment and Planning*, 4, 347-71.

27. Rivizzigno, V. L., 1975. "Individual Differences in the Cognitive Structuring of an Urban Area." In R. G. Golledge (ed.), *On Determining Cognitive Configurations of a City*, pp. 471-82.

28. Shepard, R. H., 1962. "The Analysis of Proximities: Multidimensional Scaling with an Unknown Distance Function, I," *Psychometrika*, 27, 125-40.

29. Shepard, R. H., 1962. "The Analysis of Proximities: Multidimensional Scaling with an Unknown Distance Function, II," *Psychometrika*, 27, 219-46.

30. Spector, A., 1975. "An Exercise in the Interpretation of Multidimensional Scaling Configurations: The Case of the Newcomers." In R. G. Golledge (ed.), *On Determining Cognitive Configurations of a City*. Columbus, Ohio: Department of Geography, Ohio State University Research Foundation.

31. Spence, I., 1972. "A Monte Carlo Evaluation of Three Nonmetric Multidimensional Scaling Algorithms," *Psychometrika*, 37, 323-55.

32. Spence, I., and N. W. Domoney, 1974. "Single Subject Incomplete Designs for Nonmetric Multidimensional Scaling," *Psychometrika*, 39, 469-90.

33. Spence, I., and Ogilvie, J., 1972. "Stress Values for Random Rankings in Nonmetric Multidimensional Scaling." University of Western Ontario, Research Bulletin #225.

34. Stea, D., 1976. "Program Notes on a Spatial Fugue." in G. Moore and R. G. Golledge (eds.), *Environmental Knowing*, pp. 106-20. Stroudsburg, Pa.: Dowden, Hutchinson and Ross.

35. Thrall, G., 1975. "Multidimensional Scaling Analysis of the Cognitive Structure of Columbus: The Case of the Intermediate Group." In R. G. Golledge (ed.), *On Determining Cognitive Configurations of a City*, pp. 349-62. Columbus, Ohio: Department of Geography, Ohio State University Research Foundation.

36. Tobler, W., 1976. "The Geometry of Mental Maps." In R. G. Golledge and F. Rushton (eds.), *Spatial Choice and Spatial Behavior*. Columbus, Ohio: Ohio State University Press.

37. Tolman, E. C., 1948. "Cognitive Maps in Rats and Men," *Psychological Review*, 55, 189-208.

38. Werner, H., 1957. "The Concept of Development from a Comparative and Organismic Point of View." In D. B. Harris (ed.), *The Concept of Development*, pp. 125-48. Minneapolis: University of Minnesota Press.

39. Young, F., 1968. "A FORTRAN IV Program for Nonmetric Multidimensional Scaling." Chapel Hill, N.C.: University of North Carolina, Thurstone Psychometric Laboratory, Research Report #56.

40. Zannaras, G., 1973. "An Analysis of Cognitive and Objective Characteristics of the City: Their Influence on Movements to the City Center." Columbus, Ohio: Ohio State University Research Foundation, Research Foundation Final Report #3086.

41. Zeller, R., and V. L. Rivizzigno, 1974. "Mapping Cognitive Structures of Urban Areas with Multidimensional Scaling: A Preliminary Report." Columbus, Ohio: Department of Geography, Ohio State University, Discussion Paper #42.

Chapter 3.3 **Recovering the Dimensions through Multidimensional Scaling — Remarks on Two Problems**

B. Marchand

I have used multidimensional scaling at different times to recover the geometric configuration of elements in a mental space and to compare it to an objective configuration of these elements [4]. In the process, I have run into two difficult problems, which I would like to discuss here. The problems concern the imposition of structure in the data-gathering stage and the fractional dimensionality situation.

PROBLEM 1

How to Avoid Imposing a Certain Dimensionality on the Recovered Space

Experimenter intrusion into data-gathering situations is all too common in geography. Many researchers fail to recognize that task-related structure can be imposed both by the direct intervention of the experimenter and by the procedures he or she selects. Consider an experiment using the following method: people are asked to draw a map of some landmarks in a city; then, all the possible pairs of distances are measured on those maps. After appropriate transformations, the pairs are used to recover, through MDS, the geometric configuration of the landmarks and an identification of the space that best facilitates their mapping.

Drawing the points on a piece of paper, however, introduces a strong (and uncalled for) constraint in the process—the resulting map is necessarily two-dimensional. When the subject set is large (e.g., over 100) and when all the distances are averaged over all the different surveys to produce a single "common" map, the dimensionality of the "average" map may be larger than two. Still, the bias introduced by the original constraint (i.e., using a piece of paper) may have a perturbing effect that is not clear.

I have tried to solve the problem in the following way [1]. I divided a set X of thirteen landmarks into different subsets A, B, and C, so that (1) the union of A, B, and C is X; and (2) the intersection of A, B, and C consists of two points very far apart that are used to give the unit distance common to the three subsets. The preliminary results are promising. The configuration space containing the thirteen landmarks is recovered with a very low stress, on four dimensions.

At this point, a general question can be raised. Given n landmarks (points), what is the optimal number of subsets? (Let us assume, for the sake of simplicity, that all subsets contain the same number of points; let n be the total number of points (landmarks); x the number of points per subset; NC the total number of different subsets. Then we have $n \geq x \geq 3$.) Three criteria might be used at this point. Criterion 1: We might want to maximize the information we get on a point (that is, to maximize the probability a point is included in a subset) and then maximize $P = x/n$. For n given, this is equivalent to maximizing x (i.e., taking only one subset). This conflicts with our desire to use various subsets to avoid imposing a two-dimensional solution. Criterion 2: We want to be able to obtain as many dimensions as possible; that is, we want $NC \geq 1$ (the maximum possible number of dimensions). This condition ensures there will not be any bias; it is easily met for any meaningful value of x and n. Criterion 3: This constraint is more subtle—assigning some landmarks together to a particular subset may introduce a special bias. It is clear from previous studies that people represent (or omit) specific landmarks according to some very interesting processes. For example, various restaurants may be represented or restaurants as a class may be omitted altogether; or, when two points are very close, one may be omitted and the other is assumed to represent both.

The processes set out in criterion 3 are not yet clearly known, but they are likely to bias a few among all the NC possible configurations. This could be done by decreasing the number of subsets used, but this would conflict with criterion 2. It is safer to maximize the number of possible configurations we are choosing our subsets from, that is:

$$\text{Maximize } NC = C_n^x = n!/X! \cdot (n - x)!$$

which is equivalent to maximizing a monotonically related function:

$$\text{Maximize } \log NC = \log n! - \log x! - \log(n - x)!$$

Its derivative is:

$$\frac{d\log NC}{dx} = -\log x + \log(n - x)$$

which is zero for $n - x = x$; that is:

$$x = n/2$$

This is a maximum since:

$$\frac{d^2\log NC}{dx^2} = -\frac{1}{x} - \frac{1}{(n - x)} < 0$$

In conclusion, to ask people to draw maps may force them to represent their mental concepts artificially in a two-dimensional space. Such bias could be avoided by breaking down the points into subsets, each containing approximately half of the points and all of them having, as their intersection, the same two well-chosen points.

PROBLEM 2

Spaces with a Fractional Dimension

The second problem seems to lead to fascinating theoretical developments. It appeared when I asked students to estimate the distance between two points A and B, which were estimated to be 40 minutes apart. Consider the point C, at a quarter of that distance, close to A; the distance AC should be estimated at a quarter of AB (i.e., 10 minutes), but it is often overestimated, say, 20 minutes. Choose next a point D, at a quarter of the length AC, close to A; AD is again overestimated, 10 minutes instead of 5 minutes, and so on. (See Figure 1.)

The estimation error is generally not as regular as that given in this example, but it seems pretty consistent. One might be tempted to treat it as "just an error." But a much deeper interpretation might be proposed. It seems that details keep popping up in the observer's mind as the path considered decreases in size. A logical consequence is that the distance between A and B will eventually be infinite if we imagine that the phenomenon will continue at smaller and smaller scales.

Mandelbrot [2, 3] has shown that such problems appear when we use too few dimensions to measure a distance. It is characteristic of a spatial structure exhibiting an inner homotethy. In this example, there is an homotethy $(r = \frac{1}{4})$.

$$AC = AB/4; AD = AC/4, \text{etc.}$$

Notice that at level 2, if we again form the segment AB, it will have a length of 80 minutes (i.e., eight segments of 10 minutes).

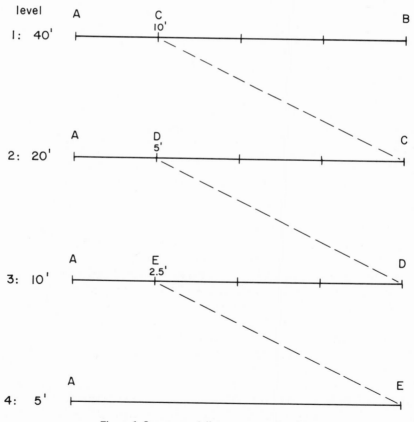

Figure 1. Sequence of distance overestimations.

In this example, the monotethy ratio is constant at each level ($r = \frac{1}{4}$), and the number of segments of level ($n + 1$) that form the distances at level n is $N = 8$. Mandelbrot proposes to define the number of dimensions of a figure considered as:

$$D = \log N / \log(1/r)$$

In the example quoted above, $D = \log 8 / \log 4 = 1.5$. In this way, the concept of a *fractional dimension* appears.

A figure of dimension D will have an infinite length when measured with a dimension $d < D$ and a length of zero when $d > D$. Only if it is measured with its right dimension D, will it present a finite length. The question arising from this example is: are we not dealing with a mental representation of fractional dimensionality, instead of simple random errors? Finding an answer to this question could point toward some interesting areas for future research.

Given that a situation such as that described above can exist, how do we measure distance in a space with fractional dimensions? The Minkowsky metric:

$$d_{AB} = [\sum_i |x_{Ai} - x_{Bi}|^p]^{1/p}$$

should become:

$$d_{AB} = [\int_o^D |x_{AD} - x_{BD}|^p f(D)^p dD]^{1/p}$$

with $f(D)$ a function of the dimension. I am not able to proceed further with the consequences of these thoughts at this time, but I am involved in correspondence with Mandelbrot and I hope, with his help, to clarify the question.

REFERENCES

1. Kooser, J., and B. Marchand. "The Perception of Urban Space—A Case Study," unpublished manuscript, Department of Geography, University of Toronto.
2. Mandelbrot, B., 1975. *Les Objets Fractals*. Paris: Flammarion.
3. Mandelbrot, B., 1975. "Physical Objects with Fractional Dimension: Seacoasts, Galaxy Clusters, Turbulence and Soap," *Bulletin of the Institute of Mathematics and Its Applications*.
4. Marchand, B., 1974. "Pedestrian Traffic Planning and the Perception of the Urban Environment," *Environment and Planning*, 6, 491-507.

Chapter 3.4 Issues in Data Collection and Reliability in Marketing Multidimensional Scaling Studies — Implications for Large Stimulus Sets

Terry Deutscher

Academicians and practitioners in the field of marketing have been quick to capitalize on the advances that psychometricians have made in the area of multidimensional scaling. In fact, it would not be presumptuous to claim that the fields of marketing and geography are probably the two that have made the greatest practical use of the techniques.

MDS applications in marketing are described in books by Green and Carmone [13] and Green and Rao [14], as well as in the journal literature in the field, most notably the *Journal of Marketing Research*. The techniques have been used for innumerable purposes ranging from product positioning and image measurement through naming and pricing of new goods. Marketing scholars, in evaluating the promise of multidimensional scaling for solving such marketing problems, have had to worry about many of the same issues that have been of concern to geographers and psychometricians. Two of these problems, scaling large stimulus sets and measuring data reliability, will be the main concern of this paper.

MULTIDIMENSIONAL SCALING AND MARKETING

To set the stage for this discussion it will first be useful to describe some of the attributes that characterize most multidimensional scaling studies in marketing. In some cases, these properties are fairly distinctive to marketing; in others, they are not. Most marketing studies employ individuals' *direct* measures of similarity, for example, by asking consumers to judge the amount of difference they see between pairs of stimuli. Using these direct similarity measures instead of *derived* ones (e.g., similarities computed from confusion matrices [21] or aggregate counts [17]) probably gives a researcher greater concern about data reliability, and it may force him or her to consider aggregating individuals [1, 2].

As Summers and MacKay [24] point out, there is another reason that marketing researchers should be very concerned about data reliability. In most cases, the stimuli employed in marketing studies (e.g., credit sources, automobiles, retail stores) are more cognitively complex than the stimuli of interest in other disciplines (e.g., colors, tones, distances). Therefore, the danger of unreliability resulting from shifts in evaluative structure from one judgment pair to another is exacerbated in marketing studies [9]. For example, suppose that a panel of consumers is requested to make judgments about automobiles at several points in time. In making a given paired similarity judgment, a person might consider the pair of cars on any combination of these attributes—economy, acceleration, reliability, appearance, prestige, handling ability, service reputation, future trade-in value, and others. In considering such complex stimuli, it is certainly possible, if not likely, that an individual's evaluative framework might vary between one time point and another, even if no true shift in his or her cognitive structure had occurred. Even worse, these variations might well occur between paired judgments made at the same point in time.

Marketing also differs from some other fields by usually not having a "true" reference base for cognitive structure. In some respects, geographers are fortunate (or, perhaps, unfortunate, depending upon one's point of view) to have physical maps of distances against which maps generated by MDS can be compared. Marketing studies usually deal with a perceptual space of brands, in one form or another, and there is no "true brand space" against which consumers' images can be compared.

Finally, marketing researchers differ somewhat from their compatriots in other fields by having greater interest in measurement over time [19]. From a practitioner's viewpoint, much of marketing strategy is aimed at bringing about a shift in consumers' perceptions and preferences [4]. For this reason, marketers are perhaps even more concerned than other researchers with making comparable measurements over time. Consequently, they *should* worry a lot about measurement reliability.

It seems clear that marketers and other MDS researchers have common enough interests that they can benefit from each others' work. Because of the nature of their research, with much of the data coming from unsophisticated consumers who cannot provide limitless amounts of high-quality data, marketers have much to gain from advances (especially in the data-collection area) in handling large stimulus sets. Of particular importance to the marketing field is the measurement of reliability. This issue should be a consideration in studies on data bases of any size, but it is critically important with larger data bases, for which fatigue and boredom might present an even more serious threat to data quality.

MDS, MARKETING RESEARCH, AND LARGE STIMULUS SETS

When faced with a problem in studying a market that contains a large number of brands (e.g., automobiles, breakfast cereals, retail stores, etc.), researchers have usually sidestepped the issue by considering only a subset of the brands. To collect complete data on a large brand set with conventional methods (e.g., magnitude estimation, ranking all possible pairs, and *n*-dimensional rank ordering) would be prohibitively expensive. Data quality would also suffer severely.

The Rao-Katz Study

Sensing the importance of this problem, Rao and Katz [20] decided to conduct a systematic evaluation of several alternatives for collecting similarities data on large stimulus sets. Since their study appeared only in the marketing literature, most readers are probably unfamiliar with it. For this reason, it will be discussed at some length in the following paragraphs.

Seven data-collection methods were advanced in the article as being suitable for large stimulus sets. They fall into three groups— three "subjective grouping" methods (in which the respondents split the stimuli into groups), two "pick" methods (in which the respondents choose a number of stimuli that they judge to be most similar to a given reference stimulus), and two "order" methods (in which the respondents rank several chosen stimuli according to their similarity to the reference stimulus).

Subjective grouping with a fixed number of groups is the first data-collection method. In this method, each respondent organizes the stimuli (n) into groups of similar objects. The number of groups (K) to be formed is specified in advance. Then, for each individual, an n x n matrix can be formed, with each cell containing a zero (if the row and column stimuli were not placed in the same group) or a one (if they were grouped together). To scale groups of people the individuals' matrices are summed. To scale individuals the n x n matrix is treated as a group of n stimulus profiles and a distance matrix between all possible pairs of stimuli is calculated.

Rao and Katz's second method is subjective grouping with variable numbers of groups. This method is the same as the first, only the choice of K (the number of groups) is left to the individual respondents.

Hierarchical subjective grouping is the third method. The subject starts by forming K groups, as in the first method. Then he or she merges the two groups that he or she considers to be the most similar. He or she continues merging groups until all stimuli are in one overall group.

In the fourth method, picking a fixed number of stimuli most similar to a reference stimulus, the subject is given a reference stimulus and then is asked to choose the K stimuli (from the remaining $n - 1$) that are most similar to the reference. This iteration is repeated $n - 1$ times with different reference stimuli. Then, a zero-one matrix (n x n) can be developed, with cells equaling one when the column stimulus is picked as being similar to the row stimulus.

Picking a variable number of stimuli most similar to a reference stimulus is the same as the fourth method, only the choice of K and the number of similar items selected at each iteration is left to the respondent.

Rao and Katz's sixth method is to order a fixed number of stimuli most similar to a reference stimulus. Data collection here is the same as for the fourth method, only in this case the respondent *ranks* the K stimuli (which he or she chose) according to their similarity to the reference item.

In the seventh method, ordering a variable number of stimuli most similar to a reference item, the choice of the number of items selected as most similar to a reference item is left up to the subject. Otherwise, this method is the same as the sixth.

The seven methods were compared by using a simulation performed for twenty generated respondents on a forty-stimulus, two-dimensional "true" map. In the methods with a fixed K (the first, fourth, and sixth), both four and eight groups were used. Where K was variable (second, fifth, and seventh methods), K was permitted a maximum value of eight. The hierarchical grouping method (three) was started with eight groups. Each of the twenty simulated individuals' "private spaces" for the forty stimuli could be fit perfectly by an INDSCAL model. In other words, the dimension-stretching factors that were applied to the two dimensions were chosen so that the sum of their squares was 1.0.

Three types of multidimensional scaling analysis were performed on the simulation data generated with the seven data-collection methods. The TORSCA algorithm was employed for scaling the data from the groups of twenty pseudoindividuals. Both metric and nonmetric methods were used. Last, the individuals' data were scaled with INDSCAL.

The vector of interpoint distances for the original configuration was correlated with the vector of distances generated from each data-collection-and-scaling combination. The results are presented in Table 1 (a condensed version of Rao and Katz's [20, p. 493] Table 6).

The subjective grouping methods typically performed worse than the pick and order methods, especially in the case of the INDSCAL model. This conclusion is perhaps not a surprising one, because the grouping methods provide somewhat less information than the others, which are also undoubtedly more time-consuming to administer.

Rao and Katz conclude that "there was no consistent superiority of the order methods over the pick methods" [20, p. 493]. Although this conclusion appears to be warranted for the case of $K = 8$ (in which the two methods produce almost equivalent results), it is more debatable for the $K = 4$ analyses. There, except for the INDSCAL model, the results for the order method appear to be much superior. Of course, in making a decision about which of the two would be preferable, one must consider that the order method is the more complicated of the two. In this method, each of the $n - 1$ sets of similar stimuli that are merely chosen in the pick method must also be ranked according to their similarity with the reference stimulus.

The bottom line in the table gives the results obtained when no economies in data collection were employed; that is, when all 780

TABLE 1
Interpoint Distance Correlation between Original and Solution Configuration

Method	Interpoint Distance Correlation		
	Group (Metric)	Group (Non-metric)	Individual Differences (Metric)
1. Subjective grouping; K = 8	0.44	0.70	0.02
1a. Subjective grouping; K = 4	0.69	0.78	0.58
2. Subjective grouping; variable K	0.71	0.81	0.39
3. Hierarchical subjective grouping	0.73	0.79	0.69
4. Pick K most similar; K = 8	0.67	0.87	0.80
4a. Pick K most similar; K = 4	0.54	0.65	0.75
5. Pick variable K; K ≤ 8	0.72	0.80	0.72
6. Order K most similar; K = 8	0.69	0.87	0.79
6a. Order K most similar; K = 4	0.66	0.86	0.75
7. Order variable K; K ≤ 8	0.66	0.85	0.71
8. Completely ranked matrices*	0.91	0.93	0.85

SOURCE: Adapted from Rao and Katz [20, p. 493].
*The data scaled for this method came from ranking the entire set of pairs of stimuli (i.e., 40 × 39/2 = 780 pairs).

pairs of stimuli had been ranked. It is interesting indeed that the recovery of the original configuration using this method was not (save for the group metric analysis) a great deal better than the "pick eight" procedure (the fourth method).

Examining Table 1 by comparing columns instead of rows is also instructive. Group scaling by the nonmetric TORSCA procedure is *always* better in terms of recovery than the metric solutions, usually by a wide margin. Furthermore, the group nonmetric approach produces a better solution than the INDSCAL method in all cases save one (fourth method, "pick four"). Even in this one instance in which the group nonmetric approach is inferior to INDSCAL, the interpoint distance correlation seems to be so far out of line with other results that one wonders whether it is perhaps a local minimum solution.

The Rao-Katz article emphasizes a contribution of the marketing literature to the scaling of large stimulus sets. It offers several approaches to the data-collection problem; and, based upon the results of the analysis, some very tentative conclusions are possible. It appears at this point that the so-called "pick" methods (first advanced by Coombs [8] with fairly large K values might be the best trade-off between maximizing information value and minimizing data-collection effort. However, before firm conclusions can be reached about the value of these procedures, the area must be investigated more fully. In the last part of this section, some potentially fruitful avenues are proposed for further research.

A simple extension of Rao and Katz's study would be well received in the literature. It would be worthwhile to investigate different levels of the number of stimuli chosen for similarity with the reference object (K in the pick and order methods). There is some evidence in the Rao-Katz research that the benefits gained from the additional data-collection effort of the order methods depend upon the level of K. It would be beneficial for researchers with large stimulus sets to know more about this relationship.

In performing modifications of Rao and Katz's work, future researchers could also incorporate some other ideas. They could easily offer informed opinions about the cost/benefit trade-offs of the different methods by conducting a pilot study that measured costs of additional data collection (for example, in ordering the stimuli selected in the pick methods). It would be useful to learn more about these trade-offs for data sets of different sizes, from fifteen or twenty up to fifty.

Future studies could incorporate the notion of error, much as the investigators of the properties of stress have [22, 23, 25, 27]. They could also check the nonmetric solutions for the presence of local minima [3]. Last, it is possible that some efficiencies might be gained by altering the choice procedure in the order methods. Currently, the stimuli chosen are the ones that are closest to the reference stimulus. By choosing only the objects *most similar* to the reference point, subjects are providing data only on the shortest distances in their perceptual configurations. Much might be gained by having them also rank a few of the stimuli that they perceived to be *most distant* from the reference.

RELIABILITY OF DATA

Hand in hand with a researcher's concern about recovering configurations of large data sets is a related issue—data reliability. Although unreliability can be a problem in data sets of any size, it is likely to be more serious in situations in which a large amount of data is collected from unsophisticated subjects during a boring, repetitive procedure. Unfortunately, this description is very appropriate for the typical marketing multidimensional scaling study.

Considering the importance of the matter, it is somewhat

surprising that more investigation of data reliability in MDS studies has not been done. As Summers and MacKay [24, p. 289] state in a recent article, "To date much of the research on MDS has focused on the technical characteristics of different MDS algorithms and relatively little attention has been given to the validity and reliability of MDS techniques in the development of perceptual maps." Many of the researchers in scaling have generated their own data in simulation studies, or else they have used derived similarities that did not require substantial amounts of data from individuals. Perhaps these are some of the reasons for the relative inactivity in the reliability area.

Weksel and Ware [26] wrote one of the few papers that considered judgment reliability in a direct similarity rating task. Correlating individual judgments in a test-retest situation, they found individual reliabilities ranging from 0.34 to 0.86, with a median of 0.71. Jain and Pinson [15] performed an experiment in which they considered whether the order of presentation of paired comparison tasks affects subjects' judgments. They also tested the effects of shifts in the level of subjects' attention to comparison tasks and shifts in the levels of subjects' commitment to their judgments. In all cases, they found no significant effects on the judgments themselves. However, no manipulation check was performed, so, as the authors acknowledge [15, p. 438], "Failure to find an attention effect and a commitment effect may be due to ineffectiveness in manipulating these factors."

MacKay has been one researcher who has recognized the importance of data reliability in evaluating MDS studies. In fact, in a recent article [18], he suggests it as one of the primary criteria for deciding whether graphic (i.e., hand-drawn) or nongraphic (MDS procedures involving pairwise data-collection techniques) methods are better for generating cognitive maps of spatially defined stimuli. As it turned out, the hand-drawn maps (of U.S. cities) were rated by the subjects as being more like actual maps. Actual comparison of the two with physical maps confirmed this judgment.

MacKay also collaborated with Summers in a recent study of the reliability of direct similarity judgments of automobiles [24]. Their measure of reliability was developed from test-retest judgments on eleven brands, taken one month apart from a group of thirty-seven students. The data were collected with the method of anchor-point ordering and instructions provided by Green and Rao [14, p. 153]. The data were first triangularized by TRICON [6]; then, they were scaled with KYST [16].

If no actual change had occurred in the respondents' perceptions (a reasonable assumption in the case of a market in which there was high brand awareness), reliability could be measured by assessing the congruence of each individual's two perceptual maps. Summers and MacKay used a number of measures of congruence: (1) Cliff's ϕ [7] —mean value 0.855; (2) product-moment correlation of the interpoint distances in the two maps—mean value 0.782, (3) cosines of angles between equivalent points in the two transformed (with Cliff's C-MATCH [7]) maps—mean value 0.822; (4) average distance separating equivalent points in the two transformed maps—mean value 0.446; (5) Spearman rank correlations of the test and retest similarity judgments as determined by TRI-CON—mean value 0.765.

Summers and MacKay [24, p. 292] report that these scores were "discouraging." It is, however, fairly difficult to evaluate the results in the abstract (i.e., to understand whether a ϕ of 0.855 or an interpoint distance measure of 0.782 represents high, medium, or low congruence). The authors thought that one reason congruence was not higher might have been that the scaling model was not appropriate. To test this hypothesis, they correlated stress (mean stress was 0.135 on the test data) and each of the congruence measures one through four. None of the correlations was significant at the .05 level, although most had the expected sign. Another possible reason for the low reliability scores could have been the fact that no context was provided for the judgments. Respondents were merely asked to give "general similarities."

The recent attention devoted to considering data reliability in MDS studies is encouraging. The article by Summers and MacKay [24] is by far the most comprehensive on this topic in the marketing literature to date. However, the method of reliability measurement employed in this research, like that of Weksel and Ware [26], involves a repeat of the entire data-collection procedure at a second point in time. As such, it is undoubtedly a very good method for studies whose focus is on *reliability itself*. However, for researchers (especially practitioners) who are primarily investigating other issues (e.g., product positioning and pricing and attitude structure), this method would be rather cumbersome. Especially when a large stimulus set is being studied, it would be a costly process to have subjects repeat their whole set of judgments. In fact, very low reliability scores might even be produced because of the respondents' boredom during their second set of judgments. It seems clear, therefore, that an alternative and less cumbersome measure of reliability would be useful for most MDS studies, in the marketing field at least.

Although in most scaling studies it would be impractical to have the respondents repeat *all* the similarity judgments, it is certainly feasible to have them repeat *some* of them. When respondents are making forty or fifty (or more) direct paired comparisons, they are unlikely to notice that a few have been repeated. Even a few data points on repeated judgments can provide a researcher with a valuable indication of the quality of his or her data.

Day, Deutscher, and Ryans [11] considered the problem of developing a suitable reliability score from these partial sets of repeated judgments. Such a measure should be comparable across data sets with different numbers of stimulus objects and different numbers of repeated paired similarity comparisons. It should also include the rank order of the *entire* set of judgments, not just the subsets that were repeated. It is the rank order of the whole set of similarities that determines the configuration resulting from non-metric scaling analysis. Thus, an appropriate reliability measure should consider not only the relative rankings of the repeated stimulus pairs, but also the ordering of these pairs relative to the ones that were not repeated.

The method developed by Day, Deutscher, and Ryans [11] is computed separately for each respondent. It is designed for studies in which paired similarities are directly assessed by the respondents. The reliability score answers the question: how far is this respondents' reliability from the worst possible reliability score he or she could have achieved by repeating the similarity judgments in question? Two rank correlation coefficients were calculated—one between the original data and the repeated data and the other between the original data and a set of data containing the most unreliable values possible for the repeated judgments.

The reliability score was determined from the formula:

$$\text{Reliability} = 1 - \frac{(1 - r_{o,r})}{(1 - r_{MIN})}$$

where $r_{o,r}$ is a Spearman rank correlation coefficient between a vector (A_1) that contains the ranking of original pairwise similarity judgments and a ranked vector (A_2) that differs only by the substitution of the repeated judgments for the appropriate originals.

r_{MIN} is a Spearman r between the original ranked vector (A_1)

and a ranked vector (A_3) with the repeated judgments assuming their worst possible (i.e., most unreliable) value.

As an illustration consider a simple example in which each respondent makes similarity judgments on all possible pairs of nine stimuli on a seven-point scale (with end points being one for almost identical and seven for totally different). Thus, the original data vector (A_1) contains thirty-six elements. Suppose that the respondents, unknown to them, also repeated their judgments on five pairs. The first step in constructing the A_3 vector is to replace in the A_1 vector each of the five original judgments that were repeated with a one and then a seven to see which value would show the larger shift in rank from the original judgment. In this way, a ranked vector representing the lowest possible reliability is found. Thus, r_{MIN} (the correlation of vectors A_1 and A_3) is the *lowest* possible correlation between the original data and the original data after substitution of five repeated judgments. The correlation using the actual results of the repeated judgments ($r_{o,r}$ between vectors A_1 and A_2) must fall between r_{MIN} and 1.0 The closer $r_{o,r}$ is to r_{MIN}, the lower the individual's reliability.

In this study [11] the authors presented reliability results from six consumer studies. The samples in these studies are generally representative of the total population or, when appropriate, the user group. A condensed description of the studies and the mean reliability results are shown in Table 2. In averaging these results across respondents, substantial individual differences are masked. For example, the average reliability of the 141 subjects in the T_2 credit survey was a reasonably respectable 0.829. However, individual reliabilities in this sample ranged from 0.161 to 1.0, with a standard deviation of 0.173. This data base will be discussed further in the next section of this paper.

DATA QUALITY AND INSTABILITY
OF PAIRED COMPARISONS

If the reliability measure described in the previous section is valid, it should certainly be related to instability of perception over time. That, essentially, is the hypothesis that this section will test.

Data from two studies were used to conduct this test; they were the two credit studies used in Table 2 in the previous section. The T_2 study consisted of reinterviews ten months later on 141 people who had completed the similarity judgments at T_1 [12]. In both studies, judgments were made in the context of obtaining credit for buying a $400 refrigerator.

TABLE 2
Reliability Results from Consumer MDS Studies

Study	Sample Size	Number of Stimuli	Data-Collection Procedure	Extent of Experience with Stimuli*	Number of Repeated Judgments†	Mean Reliability Score
1. Credit source study (T_1)	336	9	In-home personal interviews	Moderate	2	0.829
2. Credit source study (T_2)	141	9	In-home personal interviews	Moderate	2	0.853
3. Blender study (subsample A)	74	12	Laboratory-setting paper-and-pencil administration	Low	4	0.800
4. Blender study (subsample B)	114	13	Laboratory-setting paper-and-pencil administration	Low	4	0.793
5. Beer study (exploratory)	52	10	Focus group interviews	High	5	0.863
6. Beer study (field)	124	11	In-home personal interviews	High	2	0.854

*Reflects the likelihood of respondents' exposure to more than one alternative through advertising or usage.

†Day, Deutscher, and Ryans [11] suggest that in future studies involving repeated measurements at least $n/2$ judgments be repeated for stimulus sets of size n.

Measures of reliability at T_1 and T_2 were developed with the method of repeated judgments. Reliability scores for the 141 people who provided data at both time periods are presented in Table 3. There appears to be a wide distribution of scores at both time points, so a good comparison of the perceptual instability of reliable and unreliable respondents should be possible.

Making this measurement at two points in time permits a check on its consistency. If the reliability score were measuring what it was designed to, it would be reasonable to expect some congruence between the same individual's scores at T_1 and T_2. That seems to be the case in Table 3. The value of tau calculated from the data in the table is significant at the .01 level of confidence, so it would appear that this is a "reliable" measure of reliability. The measure of perceptual stability between T_1 and T_2 was the Spearman rank correlation between the vectors of paired comparisons at T_1 and T_2.

Perceptual measures were taken ten months apart. Therefore, it must be recognized that there could be two different reasons for any changes in the data from T_1 to T_2. The change score could

TABLE 3
Consistency of Reliability Scores over Time
$(Tau_b = 0.174, p < .01)$

		Reliability at T_2				Row Total
		0-0.39	0.4-0.79	0.8-0.99	1.0	
	0.-0.39	1*	1	8	0	10
	0.4-0.79	3	10	13	8	34
Reliability at T_1	0.8-0.99	1	14	41	15	71
	1.0	0	2	13	7	22
	Column Total	5	27	75	30	137

*Read as "one person in the sample had a reliability score between 0 and 0.39 at both points in time."

reflect "noise" (poor data quality), or it could result from a true perceptual change. Both studies were done during the months after truth-in-lending legislation was passed, and there is evidence [5] that some learning was still occurring among consumers.

For this reason, other variables that were thought to be potential predictors of attitude change or stability were included in the data-collection process. These variables are listed below, together with a brief description of how they were measured in the study.

1. Experience prior to the T_1 study—as measured by the number of different types of credit sources used during the five years before the first interview.
2. Credit-using experience between T_1 and T_2—as measured by a dummy variable that is one when a credit purchase was made between T_1 and T_2 and zero otherwise.
3. Education—as measured by the number of years the respondent spent in school and college.
4. Involvement—as measured by the respondent's self-reported interest in the differences among credit sources. Because this variable could not be assumed to be intervally measured, the five points on the interest scale were collapsed to a zero-one dummy variable in which the one denoted extreme interest in the differences and the zero represented the other points on the scale. This division was the one closest to the median score on the variable.

5. Confidence—as measured by the respondent's self-report of his or her confidence in making judgments on the different sources of credit. A five-point rating scale was compressed into a zero-one dummy variable in which the one represented "very confident" and "extremely confident."

6. Knowledge of credit—as measured by the respondent's ability (or inability) in the T_1 survey to give an approximately accurate estimate of the annual percentage rate of interest charged on a credit purchase of a $500 color television.

The basic question to be answered in the analysis of this data concerns the role of measurement unreliability. With the effect of the other six predictor variables held constant, is it significantly related to instability in the perceptual data? This question can be answered with a multiple regression.

The multiple regression of all seven predictor variables on perceptual stability is reported in Table 4. The predictive power of the equation is not high (the multiple correlation coefficient is .40, for an R^2 of .16), but, nonetheless, it is significant at the .01 level. To some extent, this low explanatory power can be attributed to the fact that the data were collected in two field surveys, as opposed to a controlled experiment in the laboratory.

By far the best explanatory variable in the equation is reliability. It is significant at the .01 level, so it can confidently be stated that unreliability of the perceptual data is a contributor to instability over time of the variance in the dependent variable that was explained by the regression (i.e., 16%). Reliability by itself accounts for almost half (a simple regression of reliability on perceptual stability has an R^2 of 7.4%).

With one exception, the signs of the other explanatory variables in the equation were as expected. However, none of them was significant at the .05 level, although five of the six approached that level.

And what implications do these results have for the MDS researcher? The major suggestion that emerges would be that, whenever possible, reliability checks should be included in the data-collection process for scaling research. In studies of attitude change, when much of the observed difference can really be spurious change that is attributable to measurement unreliability, this step is particularly important. In some of the worst cases, it may even be necessary to remove some subjects' responses from the analysis.

TABLE 4
Determinants of Perceptual Stability

Dependent Variable

Spearman rank correlation coefficient between T_1 and T_2 paired comparison data on credit sources

Independent Variables	Predicted Sign	B	t Statistic $(p)*$	β
Reliability	Positive	0.246	2.48(.007)	0.21
Credit experience pre-T_1 (number of types of sources previously used)	Positive	0.0182	1.27(.11)	0.117
Education (number of years)	Positive	0.0125	1.58(.06)	0.144
Credit purchase T_1 − T_2 (yes = 1)	Negative	−0.0687	1.52(.07)	−0.131
Interest (extremely interested = 1)	Positive	0.0564	1.25(.11)	0.104
Confidence (extremely confident and very confident = 1)	Positive	−0.0138	−0.33† (−)	0.028
Knowledge of credit rates (aware = 1)	Positive	0.0666	1.54(.07)	0.133
Constant		−0.048		

Regression equation: $R^2 = .160, F = 3.46$, df $= 7,127$‡; $p < .01$.

*The value of p for each coefficient refers to the probability that the null hypothesis (that the coefficient is zero) would be rejected if it were true (type 1 error).
†This coefficient is clearly not significant (for a one-tailed test it is not significant even at $p = .5$).
‡Number of residual degrees of freedom were reduced by 6 from 133 because six respondents had missing values in some of the variables that were used in the equation.

CONCLUSIONS

In a field that is as new as nonmetric multidimensional scaling, it is not surprising that different areas of application face some of the same problems. Data reliability is a topic that will be of increasing concern to users of paired similarity comparisons. The problem will be particularly acute for large stimulus sets, in which data collection problems are likely to be exaggerated.

This paper has summarized contributions from the marketing literature to the analysis of large data sets and has made suggestions for future research in the area. It has commented on the work to date in measuring reliability of similarities and has suggested a measure that is useful when large amounts of data must be collected from individuals. Baseline reliability scores were presented from several consumer studies. In a two-wave study of consumer perceptions of credit sources, reliability scores were shown to be consistent over time and to be important indications of instability when perceptual change is being studied.

REFERENCES

1. Anderson, A. B., 1970. "The Effect of Aggregation on Nonmetric Multidimensional Scaling Solutions," *Multivariate Behavioral Research, 5*, 369-73.
2. Anderson, A. B., 1973. "Brief Report: Additional Data on Distortion Due to Aggregation in Nonmetric Multidimensional Scaling," *Multivariate Behavioral Research*, 8, 519-22.
3. Arabie, P., 1973. "Concerning Monte Carlo Evaluations of Nonmetric Multidimensional Scaling Algorithms," *Psychometrika*, 38, 607-8.
4. Boyd, H. W., M. L. Ray, and E. C. Strong, 1972. "An Attitudinal Framework for Advertising Strategy," *Journal of Marketing*, 36, 27-33.
5. Brandt, W. K., G. S. Day, and T. Deutscher, 1975. "Information Disclosure and Consumer Credit Knowledge: A Longitudinal Analysis," *Journal of Consumer Affairs*, 9, 15-32.
6. Carmone, F. J., P. E. Green, and D. J. Robertson, 1968. "TRICON—An IBM 360/165 FORTRAN IV Program for the Triangularization of Conjoint Data," *Journal of Marketing Research*, 5, 219-20.
7. Cliff N., 1966. "Orthogonal Rotation to Congruence," *Psychometrika*, 31, 33-42.
8. Coombs, C. H., 1964. *A Theory of Data.* New York: John Wiley and Sons.
9. Day, G. S., 1972. "Evaluating Models of Attitude Structure," *Journal of Marketing Research*, 9, 279-86.
10. Day, G. S., and W. K. Brandt, 1974. *A Study of Consumer Credit Decisions: Implications for Present and Prospective Legislation.* Washington, D.C.: National Commission on Consumer Finance.
11. Day, G. S., T. Deutscher, and A. B. Ryans, 1976. "Data Quality, Level of Aggregation, and Nonmetric Multidimensional Scaling Solutions," *Journal of Marketing Research*, 13, 92-96.
12. Deutscher, T., 1974. *Credit Legislation Two Years Out: Awareness Changes and Behavioral Effects of Differential Awareness Levels.* Washington, D.C.: National Commission on Consumer Finance.
13. Green, P. E., and F. J. Carmone, 1970. *Multidimensional Scaling and Related Techniques in Marketing Analysis.* Boston: Allyn and Bacon.
14. Green, P. E., and V. R. Rao, 1972. *Applied Multidimensional Scaling: A Comparison of Approaches and Algorithms.* New York: Holt, Rinehart and Winston.

15. Jain, A. K., and C. Pinson, 1976. "The Effect of Order of Presentation of Similarity Judgments on Multidimensional Scaling Results: An Empirical Examination," *Journal of Marketing Research*, 13, 435-39.

16. Kruskal, J. B., F. W. Young, and J. B. Seery, 1973. "How to Use KYST, a Very Flexible Program to Do Multidimensional Scaling and Unfolding." Mimeographed report, Bell Laboratories, Murray Hill, N.J.

17. Leon, J. J., 1975. "Sex-Ethnic Marriage in Hawaii: A Nonmetric Multidimensional Analysis," *Journal of Marriage and the Family*, 37, 775-81.

18. MacKay, D. B., 1976. "The Effect of Spatial Stimuli on the Estimation of Cognitive Maps," *Geographical Analysis*, 8, 439-52.

19. Moinpour, R., J. M. McCullough, and D. L. MacLachlan, 1976. "Time Changes in Perception: A Longitudinal Application of Multidimensional Scaling," *Journal of Marketing Research*, 13, 245-53.

20. Rao, V. R., and R. Katz, 1971. "Alternative Multidimensional Scaling Methods for Large Stimulus Sets," *Journal of Marketing Research*, 8, 388-94.

21. Shepard, R. N., 1972. "Psychological Representation of Speech Sounds." In E. E. David and P. B. Denes (eds), *Human Communication: A Unified View*, pp. 67-113. New York: McGraw-Hill.

22. Sherman, C. R., 1972. "Nonmetric Multidimensional Scaling: A Monte Carlo Study of the Basic Parameters," *Psychometrika*, 37, 323-55.

23. Spence, I., and J. Gray, 1974. "The Determination of the Underlying Dimensionality of an Empirically-Obtained Matrix of Proximities," *Multivariate Behavioral Research*, 9, 331-41.

24. Summers, J. O., and D. B. MacKay, 1976. "On the Validity and Reliability of Direct Similarity Judgments," *Journal of Marketing Research*, 13, 289-95.

25. Wagenaar, W. A., and P. Padmos, 1971. "Quantitative Interpretation of Stress in Kruskal's Multidimensional Scaling Technique," *British Journal of Mathematical and Statistical Psychology*, 24, 101-10.

26. Weksel, W., and E. E. Ware, 1967. "The Reliability and Consistency of Complex Personality Judgments," *Multivariate Behavioral Research*, 2, 537-41.

27. Young, F. W., 1970. "Nonmetric Multidimensional Scaling: Recovery of Metric Information," *Psychometrika*, 35, 455-73.

Chapter 3.5 Suggestions for Identifying Sources of Unreliability in Multidimensional Scaling Analyses

Rex S. Green

A relatively unsettled issue for real-world applications of multidimensional scaling (MDS) techniques is the estimation of the reliability, or repeatability, of the estimated scale values (i.e., the dimension loadings, parameter estimates, and recovered distances) that are synonymous with the spatial coordinates, et cetera. Ideally, it would be helpful to have estimates of the reliability of: each point location, which Best, Young, and Hall [1] refer to as "wobble"; each defined scale or direction in the space that is interpreted; and the MDS solution of p dimensions. Since the issue of reliability of measurement relates to the way in which the estimated scale values will be used [5, pp. 14-29], the general topic of the reliability of MDS solutions is much too broad to be the focus of this discussion.

Let me narrow the focus thusly. How might the reliability of an MDS solution with a specified dimensionality be estimated for a single judge when the issue is repeatability of that solution for the same point in time, circumstances, and so forth? This focus bypasses the issues of fluctuations in reliability across dimensions, judges, occasions, and circumstances to address the basic concern of "how similar would these MDS results look if the observations were gathered again but independently of the first set of observations?" Presumably, the reliability for a single judge would be at the highest possible level in this case; adding more facets of generalization would be expected to lower the levels of reliability. Also, since it is likely that investigators will need parameter estimates from each judge for use elsewhere, the reliability of each judge's solution should be addressed.

There is no lack of previous effort to estimate the fit of the MDS model to the observations or some transformation of them [2, 11, 15, 16, 18]. From the assumption that the fit of the MDS model is an appropriate means of estimating the repeatability of the parameter estimates, where might there be a problem? MDS does not escape the problem of overfitting a model when there are too few observations relative to the number of parameter estimates [12, 17]. Several Monte Carlo investigations of MDS [7, 10, 13, 14, 18] have reached much the same conclusion. When there is distortion or noise present in the observations, several times (two or three) as many observations must be collected as there are parameters to estimate. Otherwise, the MDS model may track the noise, and the estimates of fit will be poor guides to estimating the reliability of the MDS solution.

Regrettably, the investigator seldom knows in advance how many dimensions, and thus parameters, are needed for each judge. Adding dimensions typically improves the fit of the MDS model, so most investigators add dimensions until the amount of improvement in fit becomes negligible. Yet, it is this very approach to selecting the number of parameters that aggravates overfitting the model. It is still true that every MDS study is at risk regarding overfitting the model, that alternative procedures are needed for selecting the dimensionality of MDS solutions, and that measures of fit tend to overestimate the reliability of the scale values.

The purpose of this paper is to stimulate the further investigation of ways to determine the reliability of MDS solutions. As a start, the following method is proposed: trisect the reliability question, construct and evaluate three indexes of reliability, and employ two of these indexes to select the number of dimensions for subsequent interpretation. Suggestions are then made for determining whether an MDS solution is minimally reliable. Some examples of how to apply this method are reported by drawing from research on the development of interactive MDS techniques, since these techniques are prone to overfitting the MDS model.

Individual differences MDS models ought to assist investigators in obtaining adequate reliability. Do they? Consider the following. The estimates of fit obtained from individual differences MDS studies depend on the sampling of judges: how many, from what populations, and what levels of measurement their observations may possess. Considerable work is needed to establish sampling procedures. How is it possible to maximize the number of judges for whom minimally reliable solutions are obtained, under the assumption that the reliability varies across judges? Does the application of an individual differences model "overfit" a few judges who tend to agree while it underfits the majority of judges who differ to some degree? Scaling larger sets of stimuli resurrects the problem of overfitting the model, given that a very small proportion of all observations are gathered from each judge. Can the measure of fit provide any indication of how repeatable the parameter estimates are for single judges since the application of the permissible transformations to a core space may not produce spatial representations that "look like" the ones based on MDS solutions for each judge?

THREE COMPONENTS OF RELIABILITY

A meaningful scaling occurs, according to Cliff [3], when these three requirements are satisfied: (1) a fairly consistent set of data relations are obtained, which are moderately numerous; (2) a mathematical model is selected, which contains variables that can be matched to the empirical conditions, and a solution procedure exists for fitting the model; and (3) some demonstration is provided of a meaningful relationship of the obtained scale to other data or scales that are external to the measurement procedure.

These requirements suggest that three components of reliability should be assessed to determine whether meaningful scaling is occurring. These three components of reliability are labeled *congruity*, *fit*, and *generalizability*.

The first component, congruity, refers to the congruence between two independent sets of observations. The data lack congruence when, for example, the judge makes a second independent rating that is very different from the first report. The congruity of the data reflects on, but is not equivalent to, the consistency of a set of data relations. That is, it is unreasonable to expect a set of moderately numerous observations to be highly consistent when repeated observations differ notably. Since consistency relates to the selection of the scaling model, whereas congruity is independent of the choice of model, it is suggested that congruity, rather than consistency, be assessed as one of the components of reliability.

The second requirement mentioned by Cliff relates to the need to assess the correspondence, or fit, of the scaling model to the observations. The fit expresses the extent to which a scaling model contains variables that match the data and yields appropriate scale values. Thus, fit relates the observations to the parameter estimates of a scaling model. These observations are the ones used to estimate the parameter values. Clearly, in order to avoid trivial solutions, the observations must be more numerous than the parameters.

The manner in which the fit expresses the repeatability of the estimated scale values can be explained by borrowing from true-score-theory terminology. The observations, or some transformation of them, are fitted by the MDS model. The fit reveals how closely the estimated scale values match the observations. Assume that there is some MDS model that, when fitted to errorless observations, will produce the true scale values. These scale values would be repeated every time a set of errorless observations is properly fitted. Thus, the perfect fit of the true scale values to errorless observations is a prediction of perfect repeatability. The repeatability of a set of estimated scale values will be higher, the more similar they are to the true scale values. However, fitting errorful observations, frequently with suboptimal models, compares two flawed sets of numbers; thus, the fit of the estimated scale values to errorful observations is only an approximation of what we need to know: the correspondence of the estimated scale values to the true scale values.

The third requirement specifies a third entity that is not a part of the fitting procedure. Referring to the congruity of the data, two sets of observations were needed to assess this component. Since one set of observations is not fitted by the model, they can provide the means for assessing the generalizability of the scale values, at least to a set of observations that is independent of the scaling procedure. The third component, generalizability, refers to the correspondence of the scale values to observations from which they could have been derived. If the model is correctly describing the lawfulness of the data relations, then the estimated scale values will match unfitted, as well as fitted, observations. Since validity refers to ascertaining the *meaning* of what is measured, repeating a set of observations that could be fitted by the model is assumed to provide information more relevant to the estimation of reliability than to validity [cf. 19, p. 396].

The major advantage of distinguishing three components of reliability is the increased ability to isolate and perhaps control the sources of unreliability. For instance, if the congruity of the data seems especially poor, improvement of data-collection efforts should be stressed. If only the fit is low, other scaling models, or alternative fitting procedures, should be explored. A low level of generalizability could be the result of overfitting the scaling model or repeating observations following a change in the data-gathering situation. Another advantage of distinguishing these components of reliability is the opportunity to make comparisons. If the fit is lower than the congruity, perhaps fitting a different MDS model will tap more of the lawfulness in the data relations. Should a model be fitted that causes the generalizability to be lower than the congruity? Such comparisons should aid investigators in finding still more satisfactory MDS results for the data collected.

Three Indexes of Reliability

Each component of reliability could be assessed in a variety of ways, with one of several coefficients. When cardinal models are fitted to ordinal data, the congruity could be measured by Kendall's tau, the fit by stress, and the generalizability by Spearman's rho. The approach being recommended, though, is selecting the appropriate product-moment (PM) correlation coefficient and assessing each component with the same type of coefficient. The appropriate PM would be Pearson's r for metric data or Spearman's rho for ordinal data. A notable gap is the fitting of observations assumed to be nominally scaled, because there is no appropriate PM correlation coefficient. The reason for taking this approach will become more clear when the inferential procedure described in a subsequent section is presented. This procedure relates PM correlations in a t-test format to ascertain the minimal reliability of an MDS solution. Another advantage accruing to this approach pertains to the optimization of measures of fit by fitting procedures. Since the correlation estimate of fit recommended here is not optimized, it may provide a more realistic estimate of the reliability of the measurements.

The indexes of reliability are based on drawing a random sample of all fitted observations, usually from thirty to fifty, collecting the corresponding unfitted observations, and comparing these two sets of observations with the scale values obtained by fitting the scaling model to all of the data. The unfitted observations must be obtained independently of the fitted observations. When the same judge provides both sets of observations, care should be taken to ensure that the judge is not aware that an observation is being repeated.

Let V contain three column vectors representing the fitted observations, the unfitted observations, and the estimated scale values, respectively. Each index is formed by correlating pairs of columns of V. The congruity:

$$\chi = \rho(v_1, v_2), \tag{1}$$

the fit:

$$\phi_p = \rho(v_1, v_3), \tag{2}$$

and the generalizability:

$$\gamma_p = \rho(v_2, v_3). \tag{3}$$

The correlation coefficient may be either Pearson's r or Spearman's rho. The subscript p refers to the pth dimension selected for fitting MDS models. The following section presents a method for determining the number of dimensions p. The reliability of each component can be assessed by examining the level of each index. The network of relations among the three indexes also bears on the reliability of scaling solutions. Both the levels and the relationships will help establish whether an MDS solution is minimally reliable.

The level of χ reflects the level of error attending the observations. It should be high enough to support fitting a scaling model, such that the estimated scale values will be meaningfully interpretable. If it is near zero, the judge may be giving a random report; this implies that there is no underlying structure to the data relations. The problem of overfitting random data, first identified by Klahr [10], can be avoided by requiring that χ be significantly higher than zero, e.g., $\chi \geq .25$, for samples of thirty to fifty observations.

When $\chi < .25$, the congruity of the data should be improved. Efforts might be made to clarify the judging instructions, select other judges, or revise the set of stimuli to increase the congruity of the data before fitting the scaling model. On the other hand, observations lacking congruence might be combined with data from other judges who have similar enough underlying structures represented in the data. A points of view or cluster analysis [3, 17] may indicate which judges can be combined. An average set of judgments can then be analyzed to seek more reliable scale values.

Since the differences among the judges can no longer be distinguished, only the reliability of an "idealized individual's" [3] solution can be estimated.

The level of fit, ϕ_p, describes the level of agreement between a sample of the fitted observations and their corresponding scale values. Of course, a better estimate of fit is available: the correlation of all the data with their corresponding scale values or recovered distances. However, ϕ_p must be based on the same sample size to be compared with the other two indexes. Some criterion level of fit should be selected, such that "clearly interpretable" solutions possess levels of fit exceeding the criterion. For instance, assume that a known multidimensional structure of p dimensions exists, that the observations contain considerable noise, and that numerous estimates of this structure are obtained. Presumably, too much noise in the data will cause the estimates of this structure to vary beyond the point at which the known structure will be recognized by the investigator. For example, a selection of major cities scattered about the United States might appear to be so jumbled that New York would be located west and south of Chicago. A criterion level of fit should help the investigator in rejecting such estimates of the known structure while encouraging the interpretation of estimates that "look enough like" the known structure. To this end a criterion for ϕ_p of .75 is proposed. The choice of this criterion level will receive some support in subsequent sections of this report. Still, some investigations of the recognizability of structural estimates are needed to improve this initial choice.

Frequently, an investigator will find that adding more dimensions will improve the level of fit. But either these dimensions become increasingly difficult to interpret or possible interpretations are weakly supported by the order of the stimuli on that dimension. The reference to "clearly interpretable" implies that the dimensionality should not be set higher than the number of interpretable dimensions just to achieve an acceptable level of fit.

The subscript p denotes the presence of several ϕs for each MDS solution. The next section describes an objective method for selecting the value of p, thence ϕ. However, the ϕ_p selected in this manner probably should serve as a starting point for deciding how many orthogonal dimensions to begin interpreting, and the criterion of interpretability should still be applied in selecting p.

The level of generalizability of the scale values, γ_p, also indicates the degree of accuracy of the MDS solution space. This index provides a more conservative estimate of reliability, since the scale values are compared with unfitted observations. Comparing this index with the levels of fit and congruity serves as a check on overfitting of the MDS model. Because most observations contain some distortion relative to the purpose of the scaling, this index should not be taken as the best single indicator of the reliability of the MDS solution. Later, a hybrid of both fit and generalizability is proposed as the best overall estimator of reliability.

Now, what would constitute a minimal criterion of generalizability? In other words, the estimation of scale values to tap the consistency of a set of data relations should not force the level of generalizability below that of the congruity. As long as the level of generalizability equals or exceeds that of the congruity, the investigator can assume that converting the data to parameter estimates has not produced an intolerable amount of distortion of the relations among the observations. Otherwise, the model or fitting procedure may need to be replaced or revised.

When the generalizability is lower than the congruity, it may be difficult to determine whether this situation is due only to sampling error or to overfitting the model to the data. Since a decision must be made, the following inferential procedure can be employed whenever the estimate of generalizability is lower than that of the congruity. Construct a correlation matrix R from V, such that the off-diagonal elements of R form estimators of the three indexes, thus, $r_{12} = \gamma_p$, $r_{13} = \chi$, and $r_{23} = \phi_p$. A reasonably powerful test of the difference between r_{12} and r_{13} can be made for $n = 30$-50 by using Hotelling's t test with William's modification, according to Dunn and Clark [6]. Note that the use of rho may somewhat weaken the power of this test. Let:

$$t = (r_{12} - r_{13}) \left[(n - 3)(r_{23} + 1)/2D_3 \right]^{1/2} /(1 + \left[(n - 3)(r_{12} + r_{13})^2 (1 - r_{23})^3 \right]/[8(n - 1)D_3])^{1/2}. \qquad (4)$$

D_3 denotes the determinant of R, which may be readily calculated from:

$$D_3 = 1 + 2r_{12} r_{13} r_{23} - r_{12}^2 - r_{13}^2 - r_{23}^2. \qquad (5)$$

A one-tailed test of the hypothesis $H_o : \gamma_p \geqslant \chi$ can be conducted using $\alpha = .05$. If the difference between γ_p and χ is significant, reject H_o; the generalizability of the model's scale values is too low, and the solution lacks a minimal level of reliability.

Table 1 gives a rough picture of the sensitivity of this test for samples of thirty-five observations. If the level of fit is about .95, then a difference of .05 tends to be sufficient to reject H_o. Once

TABLE 1
Detecting a Lack of Generalizability
for the Scaling Model

Fit

		.95	.85	.75
	−.05	if $r_{13} > .85$	if $r_{13} > .94$	if $r_{13} > .94$
$r_{12} - r_{13}$	−.10	detectable	if $r_{13} > .80$	if $r_{13} > .86$
	−.20	detectable	detectable	if $r_{13} > .25$

NOTE: $n = 35; \alpha = .05$.

the level of fit falls below .75, only very large differences can be detected. A third factor is the level of congruity. The diagonal cells of the table indicate that the level of congruity must be high enough to detect smaller differences. Only larger differences can be detected as the level of fit falls unless the level of congruity remains high. The lower right-hand cell of the table conveys why this table need not be extended. A decline in $r_{12} - r_{13}$ of .20, which is a fairly strong indication that there is a loss of generalizability taking place, can be detected as long as the levels of fit and congruity exceed their minima. Thus, setting the criterion level of fit at .75 and congruity at .25 seems consistent with the application of this test.

Selection of the Dimensionality

The difficulties associated with selecting the correct number of factors for a factor analysis can plague fitting an MDS model to a set of noisy data. To minimize these difficulties for MDS, the following procedure is described that focuses first on maximizing γ because γ is less likely than ϕ to increase as dimensions are added. The highest value of γ across dimensionalities should correspond to the best number of dimensions to retain. At times this will not be true; the highest γ_p will be associated with a ϕ_p that is too low. Then, a p should be selected that maximizes both γ and ϕ.

To select p for a particular set of data, calculate k solutions, $k \geq 5$. If the true p is expected to be 4 or higher or if the level of generalizability keeps rising, set k higher still. Determine γ_p and ϕ_p for $p = 1, 2, \ldots, k$. Round the values to two decimal places.

Select that p that maximizes γ in the fewest number of dimensions. Ignore increases of .01 and sometimes even of .02. If γ_p is too low, adjust p to more nearly maximize both γ and ϕ; trade small declines in γ for large increases in ϕ.

An Estimate of Reliability

The use of three indexes to assess three components of reliability should aid in diagnosing the sources of unreliability. Every component should possess their prescribed minimal level of reliability. Provided that they do, what, then, is the overall level of reliability? Since measures of fit tend to overestimate the repeatability of a set of parameter estimates, perhaps diluting the set of fitted observations with the sample of unfitted observations will yield more accurate estimates of reliability. In other words, before calculating the fit of all fitted observations to their corresponding scale values, replace the thirty to fifty fitted observations with the unfitted, or repeated, observations. Basing the estimate of reliability of the MDS solution on fitted *and* unfitted observations should help overcome the problem of estimating the reliability with very parsimonious samples of data.

The estimate of reliability also should describe the proportion of variance shared by the scale values and observations. It is recommended that this estimate be the square of the appropriate PM correlation [18]. Selecting a standard estimator will aid in comparing different algorithms relative to the same set of observations. Also, this estimate can be contrasted with the proportion of variance shared by the known, or true, scale values and the estimated scale values when Monte Carlo studies are performed. Note that this approach does not preclude transforming observations prior to the calculation of the coefficient.

RELIABILITY ASSESSMENT OF INTERACTIVE MDS SOLUTIONS

This section contains examples of how the calculation of these three indexes of reliability serve to establish a minimal level of reliability of MDS solutions obtained from interactively selecting an incomplete set of judgments. Young and Cliff [19] proposed the first interactive MDS algorithm—a computer program to select and present pairs of stimuli for judgment, collect the judge's responses from a computer-linked terminal, analyze these data to determine

which pairs to present next, and repeat this cycle until as few judgments as possible have been collected to reconstruct the judge's
cognitive structure adequately. The following examples are drawn
from tests being conducted on a different algorithm, one that iden-
tifies a maximum volume core, or frame, set of stimuli, following
the suggestions of Green and Bentler [9]. Green and Bentler also
supply a brief review of the mechanics of interactive scaling as well
as the rationale behind the development of a different algorithm.

Simulated Data Examples

The results of tests that were run on a preliminary version of
the volume-maximization algorithm were reported by Green [8].
A Monte Carlo technique was employed to add random distortion
to the coordinates, or true scale values, of four different configurations. Then, the distances were calculated and transformed to
nine categories, numbered 1 to 9. Additionally, thirty of these simulated judgments were identified by the interactive MDS algorithm
as requiring regeneration. In this way, the three vectors of V were
created so that the three indexes could be calculated for each of
the forty MDS solutions.

Since all results were based on known configurations and the interactive MDS algorithm checks the level of χ to decide whether
the core set of stimuli is large enough relative to the congruity of
the judgments, virtually all of these results should possess a minimal level of reliability, unless too much noise is added to the data.
In these tests either 92% or 74% of the variance of recovered distances was true variance.

First, all of the γ_p and ϕ_p were examined to select the value of p
for each MDS solution. Note that in all cases the algorithm produced results for at least as many as the known number of dimensions. Three examples of the results are provided in Table 2. In the
first example, the generalizability is maximized for one dimension,
but the fit improves notably from one to two dimensions. Select
$p = 2$, because the decline in generalizability is small relative to the
gain in fit. Remember to round to two decimal places and ignore
changes of .01 and perhaps even of .02. The second example illustrates a too-gradual rate of improvement in both fit and generalizability. Be parsimonious and select $p = 2$, which was the dimensionality of the known configuration. The third example requires the
use of some judgment. The generalizability is maximized at $p = 5$,
but the fit trickles up for $p = 5\text{-}7$. The gain in generalizability

TABLE 2
Determining the Dimensionality from the Levels of
Generalizability and Fit

	Dimensionality	Generalizability*	Fit*
Example 1	1	826-83	71
	2	806-81	91
	3	770-77	92
	4	780-78	93
Example 2	1	72	78
	2	91	96
	3	92	97
	4	92	98
Example 3	1	48	48
	2	60	61
	3	77	80
	4	89	91
	5	91	95
	6	91	96
	7	88	98

*Decimals omitted.

from $p = 4$ or $p = 5$ is only .02. Select either $p = 4$ or $p = 5$. The known dimensionality was four.

The results of applying this approach to the selection of p correctly identified the known dimensionality for 85% of the low-error MDS solutions and 60% of the high-error solutions. Overall, the known dimensionality was correctly estimated twenty-nine times; was overestimated by one, six times; and was underestimated by one, five times. Only in three instances when the pattern of values appeared unclear was a judgment call required.

Minimal levels of congruity and fit were achieved for all forty solutions; in every instance, $\chi > .25$ and $\phi_p > .75$. In fact, the lowest level of fit was .83; this reflected the fact that more than 50% of the variance was true variance. On seventeen out of the forty runs, the congruity exceeded the generalizability. For five of the runs the difference was clearly negligible. Of the twelve t tests that were performed, for only one test was the difference significant, $\alpha = .05$. Since all three indexes exceeded .95 in this case, another test was conducted, with $\alpha = .01$. The null hypothesis was not rejected. Because of the increased sensitivity of the test as the level of the coefficients increases, it is recommended that a smaller α be selected for relatively error-free results, e.g., when all indexes equal or exceed .95. Nevertheless, even with one false rejection of H_o, these results are consistent with the sampling approach.

For demonstration purposes, one of the t tests from the Monte Carlo runs is presented. Set α = .05, t_α = -1.703, with 27 df. The values of each index were r_{12} = .611, r_{13} = .663, and r_{23} = .846. The sample size was n = 30. The obtained value was t = -.6408. Thus, H_o was accepted. A difference of .052 was less than sufficient to conclude that the lower level of generalizability was not due to sampling error.

The notion of replacing fitted with unfitted observations to obtain a single estimate of the reliability was developed after these Monte Carlo tests were run. It was noted from reviewing these results that the correlation between all selected judgments (some were not fitted) and the corresponding scale values provided the closest approximation to the correlation of the estimated scale values with the known ones. The correspondence of only fitted observations tended to be too high; the sample assessment of fit, ϕ_p, tended to be more erratic, as well as to be an overestimate. Since the simulated results were not saved, the estimates of reliability could not be calculated following this proposed method.

Applications of MDS to Psychotherapeutic Assessment

Now for a look at the reliability assessment of MDS solutions for actual data. Prof. Lydia Temoshok, in collaboration with this author, is in the process of creating a list of stimuli for MDS analysis that represent key figures and events in an individual's life. Such a list can provide periodic mappings of a patient's cognitive structure, and it could be used to evaluate the course of his or her psychotherapy. Several people have interacted with the interactive MDS algorithm to provide preliminary feedback for developing this list of stimuli. Two examples of these interactive MDS results were selected to illustrate how the reliability of MDS solutions is determined.

One minimally reliable MDS solution contained four dimensions. The interactive MDS algorithm selected thirty fitted observations for collection a second time. The congruity of the data was .891, well above .25. The algorithm supplied a five-dimensional, metric solution that was pared to four, since: ϕ_1 = .76, ϕ_2 = .78, ϕ_3 = .85, ϕ_4 = .91, and ϕ_5 = .90 and γ_1 = .65, γ_2 = .66, γ_3 = .79, γ_4 = .85, and

γ_5 = .83. Clearly, both the fit and the generalizability were greatest for four dimensions. The fit of .91 exceeded the minimum of .75. The decline in generalizability, –.038, was tested by using α = .05, df = 27; since t = –1.04 and t_α = –1.70, the null hypothesis of no difference was not rejected. No estimate of the overall level of reliability was reported because the MDS algorithm has not been programmed to implement the suggestions contained herein. In this example, the fit of .91 and the generalizability of .85 are reasonably close. Thus, the overall reliability is presumed to be the square of about .88, or .77.

Another set of results did not possess the minimal level of generalizability. The values calculated for the three indexes, based on thirty-three repeated judgments, were: ϕ_1 = .47, ϕ_2 = .50, ϕ_3 = .77, ϕ_4 = .84, ϕ_5 = .83, and ϕ_6 = .84; γ_1 = .44, γ_2 = .39, γ_3 = .60, γ_4 = .64, γ_5 = .68, and γ_6 = .67; and χ = .80. Testing the discrepancy between γ_5 and χ produced t = –1.87; t_{05} = –1.70; df = 30. The lack of generalizability for this MDS solution was traced to a judging discrepancy. The judge inadvertently reversed the poles of the dissimilarity scale for several key judgments. This problem is common and can be easily handled by placing a replica of the correct scale on or above the terminal's keyboard. Failure to do so in this instance forced the loss of these results; the interactive selection algorithm had been misled into producing a suboptimal set of incomplete data. At best, the erroneous judgments could be replaced and a noninteractive MDS analysis attempted.

REVIEW

The problem stated initially was how to determine whether an MDS solution is reliable enough when the number of observations are few relative to the number of parameters being estimated—a situation that is frequently encountered when large sets of stimuli (or too few stimuli) are scaled. Two arguments against relying solely on measures of fit were presented. The suggestion was made that by randomly selecting and gathering from thirty to fifty judgments a second time, three components of reliability could be examined. Not only may a more accurate assessment of reliability be possible, but the source of unreliability may be identified as well.

Three indexes were defined for empirically assessing each component: congruity, or the correspondence of two sets of observations, only one of which is fitted by the scaling model; fit, or the degree of association between the estimated scale values and the fitted observations that correspond to the sample of repeated observations; and generalizability, or how well the scale values predict the unfitted observations. Because fit and generalizability vary over the number of dimensions being fitted, an approach to selecting the dimensionality was developed. This approach stresses finding the fewest dimensions that maximize both indexes. Nevertheless, the number of interpretable dimensions that span the space should be the ultimate determinant of the dimensionality.

Examples of the application of these methods were drawn from a preliminary Monte Carlo study of a revised interactive MDS model and studies of applications of interactive MDS to psychotherapeutic assessment. The results from the Monte Carlo simulation of judgments were consistent with the use of these indexes and the proposed criteria for establishing a minimal level of reliability. They also revealed that the best estimate of the correlation of a known solution to an estimated solution was the correlation of the estimated scale values to both fitted and unfitted observations. Application of this three-way approach to estimating the reliability for actual data collected to develop psychotherapeutic assessment procedures showed some promise, too. One example met all the criteria for a minimal level of reliability. A second example failed to achieve minimal generalizability; the source of unreliability was traced to the temporary reversal of the similarity-dissimilarity poles of the judgment scale.

Certainly, more work is needed to establish the appropriateness and usefulness of examining three components of reliability. Some of the issues still to be addressed include developing an unbiased estimate of the overall reliability of an MDS solution space; applying the three-component approach to the fitting of noninteractive MDS models, such as the cyclic design selection of pair comparisons or individual differences models; and verifying that the sources of unreliability for both simulated and actual data can be accurately traced. Despite this formidable list, the three-way assessment of the reliability of MDS solutions should begin to enhance the use of MDS in real-world applications as well as to stimulate a fresh look at the reliability of MDS solutions.

REFERENCES

1. Best, A. M., III, F. W. Young, and R. G. Hall, 1979. "On the Precision of a Euclidean Structure," *Psychometrika*, 44, 395-408.
2. Carroll, J. D., and J. J. Chang, 1970. "Analysis of Individual Differences in Multidimensional Scaling via an N-Way Generalization of (Eckart-Young) Decomposition," *Psychometrika*, 35, 238-319.
3. Cliff, N., 1968. "The Idealized Individual Interpretation of Individual Differences in Multidimensional Scaling," *Psychometrika*, 33, 225-32.
4. Cliff, N., 1973. "Scaling," *Annual Review of Psychology*, 24, 473-506.
5. Cronbach, L. J., G. C. Gleser, H. Nanda, and N. Rajaratnam, 1972. *The Dependability of Behavioral Measurements: Theory of Generalizability for Scores and Profiles*. New York: John Wiley and Sons.
6. Dunn, O. J., and V. Clark, 1971. "Comparison of Tests of the Equality of Dependent Correlation Coefficients," *Journal of the American Statistical Association*, 66, 904-8.
7. Girard, R., and N. Cliff, 1976. "A Monte Carlo Evaluation of Interactive Multidimensional Scaling," *Psychometrika*, 41, 43-64.
8. Green, R. S., 1977. "Evaluating a New Algorithm for Interactive MDS: PERSCAL." Paper presented at the meeting of the Psychometric Society, University of North Carolina, Chapel Hill, N.C., June 1977.
9. Green, R. S., and P. M. Bentler, 1979. "Improving the Efficiency and Effectiveness of Interactively Selected MDS Data Designs," *Psychometrika*, 44, 115-20.
10. Klahr, D., 1969. "A Monte Carlo Investigation of the Statistical Significance of Kruskal's Nonmetric Scaling Procedure," *Psychometrika*, 34, 319-30.
11. Kruskal, J. B., 1964. "Nonmetric Multidimensional Scaling," *Psychometrika*, 29, 1-27, 115-29.
12. Ramsey, J. O., 1969. "Some Statistical Considerations in Multidimensional Scaling," *Psychometrika*, 34, 167-82.
13. Sherman, C. R., 1972. "Nonmetric Multidimensional Scaling: A Monte Carlo Study of the Basic Parameters," *Psychometrika*, 37, 323-55.
14. Spence, I., and D. W. Domoney, 1974. "Single Subject Incomplete Designs for Nonmetric Multidimensional Scaling," *Psychometrika*, 39, 469-90.
15. Takane, Y., F. W. Young, and J. de Leeuw, 1977. "Nonmetric Individual Differences Multidimensional Scaling: An Alternating Least Squares Method with Optimal Scaling Features," *Psychometrika*, 42, 7-67.
16. Torgerson, W. S., 1958. *Theory and Methods of Scaling*. New York: John Wiley and Sons.
17. Tucker, L. R., and S. Messick, 1963. "An Individual Differences Model for Multidimensional Scaling," *Psychometrika*, 28, 333-67.
18. Young, F. W., 1970. "Nonmetric Multidimensional Scaling: Recovery of Metric Information," *Psychometrika*, 35, 455-74.
19. Young, F. W., and N. Cliff, 1972. "Interactive Scaling with individual Subjects," *Psychometrika*, 37, 385-415.

Index

Reginald G. Golledge is professor and chairman of the geography department at the University of California, Santa Barbara. He is co-editor of the journal *Urban Geography*, serves on the editorial boards of *Geographical Analysis* and *The Professional Geographer*, and is a member of the editorial review board for *Environment and Behavior*.

John N. Rayner is professor and chairman of the geography department at Ohio State University, and he serves on the editorial boards of *Physical Geography* and *Geographic Analysis*.